A Weather Legacy:
Half-Century on the Job

BOB RIGGIO
METEOROLOGIST

Bob Riggio is a member of the American Meteorological Society.

Copyright © 2020 Bob Riggio. Unless otherwise indicated, all material on these pages are copyrighted by Bob Riggio. All rights reserved. No part of these pages, either text or image, may be used for any purpose other than personal use. Therefore, reproduction, modification, storage in a retrieval system or retransmission, in any form or by any means, electronic, mechanical, or otherwise, for reasons other than personal use, is strictly prohibited without prior written permission.

Bob Riggio, e-mail: riggio@rap.midco.net.

Cover photo of fog settling across the Black Hills and the back cover lower photo of cloud-to-ground lightning in the South Dakota grasslands by South Dakota Department of Tourism.

Back cover upper photo of the author in the NewsCenter1 studio by NewsCenter1.

Cover and internal design © 2020 by Chicken Creek Communications, LLC. Printed by Simpson's Printing, Rapid City, SD. All rights reserved. Distributed by Bob Riggio. Contact the author at riggio@rap.midco.net.

Published by Chicken Creek Communications, LLC, Rapid City, SD, First Edition, 2020.

Copyright 2020
Library of Congress Cataloging-in-Publication Data
Names: Riggio, Bob, author.
Title: A Weather Legacy: Half-Century on the Job
Description: Chicken Creek Communications, Rapid City, SD, 2020.
Identifiers: Library of Congress Control Number: 2020914193| ISBN 978-0-9973849-4-9 (HC)|
Subjects: Author: Riggio, Bob | Subject: Riggio, Bob | Subject: Meteorologists – Biography – Popular works. | Subject: Television weathercasters – Biography – Popular works. | Subject: Weather – Popular works. | Subject: Meteorology – Popular works. | Subject: Weather forecasting – United States – Popular works. | Subject: Black Hills (S.D. and Wyo.) – Climate – Popular works.
Classification: LCC QC983.Rig 2020 | DDC 551.69 Rig – dc23
Summary: My thoughts on the weather found in this book, even though inspired by South Dakota's ever-changing weather conditions, apply directly to any place in the country, and anywhere in the world. Whether you reside in the Northern Plains or not, reading this book will leave you with a better understanding of this utterly fascinating, highly dynamic, and in-your-face phenomena known as weather.

ACKNOWLEDGMENT

The author wishes to acknowledge the considerable encouragement from Thomas Griffith whose guidance led me around the many road bumps on this book-writing journey. The author acknowledges the insight and expertise of Dr. Matt Bunkers, with the Rapid City National Weather Service, and meteorologists Jim Miller and David Carpenter who both reviewed and wisely critiqued the early draft of the book and made noteworthy suggestions regarding this incredible, fascinating and all too humbling science of meteorology. The author recognizes Orville Hengen, Jr., Gloria Riherd and Eric Green who help make the book totally understandable and readable to the person on the street with an interest in weather and, just as important, who located and replaced the missing commas, tiny pronouns and adverbs. And finally, the author truly values the amazing talent of Dan Lutz and Meg Roman, whose creative illustrations are sprinkled throughout the book.

DEDICATION

This book is dedicated to the late Don Larner — a Texas Aggie, agronomist, cattle rancher, golfer, gardener/beekeeper, mentor and most importantly a dear friend. A man who conveyed to me his thoughts about life, not the least of which is to cherish the good times and to adjust oneself during the not so good times. I so very much miss his understanding, counsel and friendship, even to this day.

Don Larner with the author

DISCLAIMER

All Google and Yahoo Images illustrated in this book were carefully filtered to include only those Google Images or Yahoo images labeled as "free to use or share even commercially." All other illustrations, not created by the author, have the illustrator properly credited.

TABLE OF CONTENTS

- **ABOUT THE AUTHOR** ix
- **REVIEWS** xi
- **PROLOGUE** xiii
- **WEATHER'S BEGINNINGS** 1
  - Ol' Sol – Weather's Unmoved Mover 1
  - The Angle of the Sun Causes Wide Temperature Variations 5
  - Blemishes on the Sun Produce Breathtaking Displays on Earth 6
  - With Weather, Moon a World Apart 8
  - Cycling on the Moon 9
  - A Once in a Lifetime Moon Glow 11
  - Gravity: Are We in Grave Danger? 12
- **THE ATMOSPHERE** 15
  - Weather Occurs in a Sphere of Gas 15
  - It's Cold Up There, Icarus 17
  - The Air's Gases and Role They Play 18
  - The Strange and Wonderful Properties of Water 19
  - Air in Atmosphere Behaves Like Water in a Creek 22
  - The Waters Define the Equator 24
  - The Doldrums, No Place for Man or Beast 26
  - Nighttime Luminous Clouds Light Up Sky 27
  - Badlands Upside-Down 28
  - Some Ozone are Good, Some are Downright Ugly 31
  - Air Quality Concerns Go Beyond Just Ozone 33
  - Dust Simply a Fact of Life 34
  - Even Blue Sky has Science Behind It 36
  - Morning Air Temperature Traps Plant Smoke 37
  - Fires and the Bermuda High Take the Blue Away 38

🍃 THE AIR PRESSURE ALL AROUND 41

    The Highs and Lows of Pressure 42
    Thick Air Masses Key to a White Winter 44
    Pressure on at Turkey Time 46
    Downhill All the Way 47
    East Meets West in Foggy Gap 48

🍃 THE TEMPERATURE WE FEEL 51

    Extremes Work to Make Weather What It Is 52
    Can't Stand Heat? Hydrate 53
    Snow's Complex Structure Gives it a High Albedo 55
    Frost – Another Dawn Delight 56
    Heat Wave Not Alarming Episode 58
    Raindrops Keep Falling on My Roof and Forecast 60
    Shedding Light on Wind's Chilling Effect 61
    Wind Chill Genesis 62
    When Cold Means Cold 64

🍃 THE WIND WE SENSE 67

    Jelly Sandwich Explains Wind 67
    Jelly Sandwich Law in the Third Dimension 68
    Weather, Thar' She Blows 69
    Appreciate the Northwestern Wind 70
    Northwest Winds Dominate in the Northern Plains 72
    The Black Hills Push Wind Speed 73
    Upper-Level Low Makes for South Dakota Wind 74
    Jet Stream Drives Winter Weather 75
    Wind Takes on Many Different Roles 77
    A Windy Perspective from On Top 78
    'Snow Eater' Winds Push Area Toward Warmth 80
    Snowmelt from Chinook Winds a Sublime Process 82

🍃 PRECIPITATION & THUNDERSHOWERS 85

    What is Normal Precipitation? 86
    Nature Has Own Sprinkler Track 88

Rain and Mud Follow in Fire's Path ... 89
Jasper Fires Create Rain ... 90
Some Names Pack Punch· ... 92
Nature's Own Light Show ... 93
The Genesis of Thunderstorms ... 94
Thunderstorm Adolescence ... 96
Ups, Downs of Thunderstorm Maturity ... 97
Straight-Line Winds Blamed for Damage ... 98
The Birth of a Tornado ... 99
The '98 Tempest Over Spencer ... 102
Man's Look Inside a 1951 Tornado Recounted ... 103
No Time for 'Chicken Little' ... 105
Hail, South Dakota ... 106
Severe Weather Watches Lacking for Hills ... 107
Gravity of Moving Water Heavy with Old Lessons ... 110

MEASURING THE WEATHER ... 113

A Forecast is Only as Good as its Beginning ... 113
Tiniest Variables Can Lead to Drastic Weather Changes ... 115
There's Order Amid the Chaos ... 117
Nature Knows When Bad Weather is Brewing ... 118
Cloud Predictors of State of Mind and Weather ... 120
Raindrops and Bug Splats Gauge Weather ... 121
Technological Eyes in the Sky Follow Weather ... 122
Warm Up to the Idea of Change ... 124
Science Says Climate Change is for Real ... 125
History Suggests Dry Trend ... 126
Up, Up and Away with Balloons ... 128
Santa Knows Weather ... 130

FORECASTING THE WEATHER ... 133

Weather Definitions Begin on the Front Lines ... 134
Forecasting Relies Upon Art and Science ... 135
Weather Watchers Hang Out with Models ... 137

Computers Crunch Weather Models 139
Weather: The Disorder of Things 140
Weather Forecasts Remain an Imperfect Science 142
Seat-Of-The-Pants Flying and Forecasting 143
Numbers are the Language of the Sciences 144
Keeping Weather Real 145
Drought Causes Explained 146
Drought: It's About Timing 148
It's A Matter of Perspective 150
Football and the Art of Weather Forecasting 151
When to Cancel Outdoor Activities 153
Weather Forecasts at Bat 154
A Weather Forecast Cannot be Right Everywhere 155
The Ups and Downs of Weather Forecasting 156
Let the Oceans Speak 158
El Niño/La Niña Weather or Not 159
Weather Futures? Bet on El Niño 161
Weather Forecast Predicts Cost of Power 162
Uncertainty Part of Flood Forecasting 164
NOAA Gets Weather Out Fast 165
Weather Service Not Liable for Forecasts 167

❧ THE SEASONS 169

Fall Arrives at Once 170
Numbers Say Fall Warmth is Normal 171
Why Autumn Leaves? 172
Ah Autumn, Furniture Sales 173
Conditions Gang Up in Spring 175
Summer's Heat from the Dogs 176
Weather Theory Well-Grounded 177
Waiting on Old Man Winter 178
Invasion of the Mutant Snowflakes 179
What If Humans Napped Winter Away? 181

Why Salt and No Pepper? 182
Winter's Unusual Partially the Usual 183
Other Grass Always Greener — Or Whiter 186
The Temps Might be Low but it's No Ice Age 187
How Cold was the Ninth Coldest Winter? 189
With A Cold and Snowy Winter Come the Floods 191
A Poet's View of Winter Solstice 192

PAST WEATHER EVENTS 195
War and Rainmaking 195
A Dynamite Rainmaking Theory 196
Meteorologists Make Rainmaking Discovery 198
Behind the D-Day Weather Forecast 199
Dust Bowl Drought: Nation's Greatest Weather Disaster 204
A Blizzard by Any Other Name 206
The Tale of Two Killer Blizzards 207
School Children's Storm' Ferocious, Forgotten 208
Holy Week's Blizzard a Big Pain 210
Atlas Failed, News Prevailed 212
Flood in '72 Fit Like a Puzzle 213
Nature's Clock Ticks Without the Y2K Bug 215
Time Marches On, and Back 216
Time Marches On, Deux 218
Weather an Inauguration Wild Card 219
Winter's Unexpected Beastly Savagery 221
I'm Sorry After All These Years 222

EPILOGUE 224
REFERENCES 225
GLOSSARY 226
INDEX 239
ADDITIONAL INFORMATION 250

🍃 ABOUT THE AUTHOR

Bob Riggio worked and continues to work in the field of meteorology now going on more than 50 years. He graduated from Coe College with a Bachelor of Arts degree in mathematics, Pennsylvania State University with a Bachelor of Science degree in meteorology and from the South Dakota School of Mines and Technology with a Master of Science degree in meteorology.

The author — Bob Riggio

He's plotted and analyzed numerous synoptic and local weather charts, while serving in the U.S. Air Force, and evaluated the state-of-the-art computer-generated weather models to forecast the weather for Air Force and Navy pilots, government officials and the public. Bob conducted cloud and rainfall research and studied weather's impact on air quality, authored numerous refereed papers addressing weather modification, droughts, hydrology, weather forecasting, and atmospheric pollution, and presented those papers at national and international weather and weather-related conferences. While holding the American Meteorological Society (AMS) Television Seal of Approval, he prepared and presented weathercasts for television and given numerous public speaking engagements on various weather topics to local officials, civic groups, schools, and industry. As an AMS Certified Consulting Meteorologist, Bob has given expert advice and testimony in Texas and South Dakota to attorneys for weather-related litigation purposes. Also, he consulted with universities regarding weather-related research projects and provided operational site-specific temperature, and rainfall weather forecasts to power generating utilities and to film producers.

Bob's racked up a whole bunch of weather-related experience over the past half-century while on the job. He believes that because of that extensive involvement in meteorology, he knows weather. Not based on exhaustive Ph.D. studies, but based mostly on down and dirty, trial and error grunt work, one gets from good old-fashioned long-term experience. This book offers up several of his thoughts gleaned from the many years working in this remarkable and thought-provoking field.

He would like the readers of this book to fear not what weather brings to our doorstep but to understand better and appreciate what weather gives us. Stop sensationalizing strong winter storms by labeling them "bomb cyclones," when in fact they bring life-giving precipitation. Stop voicing alarm by crying wolf when a thunderstorm initially develops on the horizon, let's at least take a moment first to appreciate its majesty. Life alone drops too many anxious experiences on our lap, and weather should not be high on that list.

So, the bottom line to the question — why read this book? Bob's answer is the following: "Our lives are not consumed by weather, but like it or not, weather consumes our lives, so treasure it. And for that reason alone, you should know what makes weather tick."

And finally, on a more personal note, Bob's been married for over 45 years and has three children and six grandchildren. He loves the outdoors in and around the beautiful Black Hills of South Dakota riding his motorcycle, fly fishing, hiking, playing golf, playing tennis and skiing — all of this, of course, weather permitting.

REVIEWS

Bob Riggio's book, *A Weather Legacy: Half-Century on the Job*, represents quite a wide range of experience from this meteorologist, and forecaster. Some of this experience shows through in Bob's description of common weather patterns across the Black Hills and Western South Dakota, although other places are mentioned too. He gives an interesting perspective on the D-Day weather and the forecasts leading up to the invasion by Allied troops. Using Bob's own educational history, he introduces meteorology for the layperson that covers a variety of topics including the atmosphere, air pressure, temperature, wind, precipitation, and weather forecasting (among other things).

Bob shifts gears from using a serious tone to humor to convey certain weather principles. He also makes use of metaphors, stories, and anecdotes to help explain complex atmospheric phenomena (e.g., the idea that the atmosphere acts as a pressure cooker that can put a "lid" on thunderstorm development). Even though Bob never worked for the National Weather Service, he has had a great partnership with them, and explains some of the National Weather Service procedures, including definitions and actions required for weather watches and warnings. Finally, scattered throughout the book are some reasons why forecasts are imperfect, which is not meant to be an excuse for wrong forecasts, but rather to point out that there are limitations to our ability to predict the weather.

In summary, this book would be useful for anyone moving to the Western South Dakota area, as well as to anyone wanting to gain a basic background in meteorology. The occasional stories and light-hearted humor make the book easy to read.

— Matthew Bunkers, Ph.D.,
Certified Consulting Meteorologist, 11 July 2018

I've never met a meteorologist who entered the field because they needed to do something for a living. It's truly a calling and always a challenge. In more than 40 years being around meteorologists, all of them passionate, Bob Riggio is the one who sets the bar for others to reach. Not only does he forecast in the most unpredictable region of the country, Bob not only makes this complex science easy to understand, but he does it through his own personal experiences and how the various meteorological events affect him. This is a great read that may inspire young readers to follow in Bob's footsteps.

— Eric Greene, News Anchor, 4 April 2019

I found Mr. Riggio's book on meteorology to be very informative with a vast knowledge of weather patterns in the Midwest region of the United States, or more specifically, the Black Hills of South Dakota. This book was easy to read and understand because of the many stories, examples, and anecdotes that were used to illustrate scientific weather principles. With Bob's weather-related experience over the past half-century, this book is a valuable resource for a basic understanding of weather, then and now.

— Gloria Riherd, Retired Elementary School Teacher

Fifty years — Bob Riggio's life adventure in Atmospheric Science is well written and covers a multitude of meteorological topics. His experiences forecasting in the State of Texas and in the beautiful region around the Black Hills of South Dakota are reflected in terms easily understood by the general public. Bob's explanations of meteorological events are non-mathematical and presented in an entertaining manner. I recommend his book for everyone having an interest in our everyday weather phenomena.

— J. R. Miller, Emeritus Assoc. Prof. of Atmospheric Sciences, South Dakota School of Mines and Technology.

Bob's collection of informative, entertaining weather essays will create a spark of recognition in readers as they absorb the colorful explanations of what they see going on in the atmosphere day by day. Bob interprets the signs in the sky in a way that everyone can understand and appreciate. His personal touch helps the reader embrace the science and enjoy the experience of reading this book.

— David Carpenter
NWS Meteorologist In Charge Retired

🍃 PROLOGUE

I believe weather uses all five of our senses to mold and lock into memory many of life's events. Weather not only acts as a backdrop but often actually shapes and guides our life's most critical life-changing encounters.

I began my career in meteorology in the U.S. Air Force, as a forecaster at Barksdale Air Force Base in Louisiana. One very foggy morning, a man wearing blue jeans and a t-shirt came into the Base Operations Office carrying two fishing poles and a plastic sack of freshly caught bass. He wanted a weather briefing to fly his T-38 jet back to Houston, TX. I told him that, due to the current zero cloud ceiling and zero visibility (WOXOF), he should delay his takeoff until early afternoon when the fog cleared. He shrugged, I checked off his "weather briefing box," and he disappeared into the morning fog. Against my advice, he started his engines and was airborne within 15 minutes.

I later learned that the pilot was none other than Apollo 13 astronaut Jack Swigert. Apparently, Swigert figured that after getting back from the dark side of the moon in the crippled lunar modular/lander, flying blind in a little fog was no big deal. One life's lesson – it's possible to do what you believe you cannot do, whether returning from the moon in a broken space capsule or taking off in WOXOF weather conditions.

Following my five-year stint in the Air Force, I earned a master's degree in meteorology from the South Dakota School of Mines and Technology in Rapid City, SD. My most vivid memory from those days was the June 9, 1972, flood. That evening, while driving with friends to have dinner, I remember commenting about the huge stationary black cloud over the Black Hills. That cloud would eventually dump more than 12 inches of rain in six hours on the Rapid Creek watershed. Later that evening, I started home after leaving my friend's house only to make it as far as the Gap (between Cowboy Hill and Skyline Drive) where the in-your-face power and fury of moving water greeted me as Rapid Creek overflowing its banks caused widespread devastation. There, near the Gap, I witnessed buildings on fire and streets turned into raging rivers of death and destruction.

The following morning, I witnessed cars stacked in a line like playing cards on a gaming table, two-story homes removed from their foundations and sitting in the middle of Jackson Boulevard, and steel railroad ties bowed by the tremendous force of the floodwaters.

Cars stacked like cards the day after the 1972 Rapid City, SD flood (Rapid City Journal)

Worst of all, thousands lost their homes and businesses, and 238 citizens of the Black Hills area lost their lives that evening. Those images imprinted in my mind the overwhelming power weather can, at times, unleash on humanity.

I met and married my wife while in Rapid City. Again, the weather played an integral part of that memory, since we planned an outdoor October wedding at the "Chapel In The Hills." The day before the ceremony greeted us with a chilly north wind, overcast skies, and rain. But, as luck would have it, the skies cleared just in time for our wedding the following day. The deep blue sky and golden aspens in full-color change gave us a crisp backdrop for great wedding photos behind the Stav Kirke Chapel and a lasting memory.

If you stop to think about some of the more significant events of your life, like those few I alluded to above, the weather may have played a sensory memory-burning role. Even many years or decades later, the weather on one given day in the present may cause you to bring back a heretofore long-forgotten memory. A memory that just may bring a smile to your lips or a tear down your cheek.

Many of my thoughts about weather presented here in these pages come from my life experiences in the Northern Plains and more specifically in and around the beautiful Black Hills of South Dakota. Here I found the weather in all its many temperaments — the good, the bad and the ugly. Some of weather's dispositions include the spring's fury, the summer's heat, the autumn's beauty, and the winter's cold.

The Black Hills area brims with many beautiful sites and diverse activities, such as Mount Rushmore, the Badlands, the natural caves, the Black Hills themselves, the prairies, canyons, rivers, and streams. It is all quite breathtaking. But there is one other local environmental attraction not to be overlooked, and that's the weather with all its wonder, beauty, power and diversity.

Here in the Black Hills, you can enjoy a panorama of clouds across the sky's natural amphitheater of azure. At one moment, an elephant appears in the clouds above, next it turns into a dragon. Other days, you can watch fair-weather cumulus clouds grow into towering cumulus clouds, then into thunderheads. There is nothing more spectacular than watching lightning illuminate an isolated storm cloud in the distance over the Black Hills just before sunset.

Badlands National Park, SD

I like to compare Black Hills' area weather to the sounds of a virtuous symphonic performance. I am not a music aficionado, but I know what I like when it comes to classical music. In my opinion, an essential element to any exemplary musical composition must be the variation of harmonious presentation. A good concerto should carry the listener from the serene woodwinds to the boisterous horns, overcome the melancholy bassoons with joyous string instruments, and rise above the relaxing oboes to the exciting timpani.

The weather musical notes that permeate our Black Hills' atmosphere not only offer up variations in audio excitement as thunder rumbles across a darkening sky, but also visual and emotional stimulation, one experiences while driving through Spearfish Canyon. Consonant with the oscillation of the string and wind instruments sending out waves

Spearfish Canyon, SD

of fluctuating musical notes, I find fluctuations in the jet stream, propagating waves of storm systems, fronts and cyclones bringing often exhilarating weather scenarios to the Black Hills.

These jet stream disturbances can abruptly transform a quiet, dry and calm afternoon into a thunderous, gusty and wet evening. They often convert a winter's morning accentuated with sunny skies and mild temperatures into a cloudy, blustery and frigid afternoon. These natural undulations of the jet stream can amazingly produce a delightful 50-degree day in Lead while dropping temperatures to below zero in Deadwood located just a mile and half down the road.

Just as the act of a bow stroking violin strings emits musical modes of oscillation, so does this vast North American landmass pluck the taut jet stream discharging these often-extreme variations to our day-to-day weather patterns. The many hundreds of miles of plains and mountains stretching between the Pacific Ocean and the Northern Plains to the west and the Gulf of Mexico to the south deviate, push and prod the shape of the jet stream permeating its symphonic weather variations from high in the atmosphere toward its Earth-bound audience. So, throw into the mix a few thousand miles of grassland, meadow and mountain ranges to pluck the strings of the jet stream, and sit back and enjoy the complete symphonic composition of our weather, here in the Black Hills.

The Black Hills' area has all the ingredients for an up-close and personal visual, audio, touch, smell, and even taste experience with weather. Where else on Earth can you sit so quietly, enjoying such an ideal blue background, and witness such an assortment of weather? My thoughts on the weather found in this book, even though inspired by South Dakota's ever-changing weather conditions, apply directly to any place in the country, and anywhere in the world.

This dynamic nature of past and present weather keeps it among the leading topics of discussion at home, in coffee shops, on social media, or in the office around the water cooler. It's my hope that this book, highlighting my thoughts on weather conjured up over the fifty years of studying, observing, plotting, forecasting, researching, broadcasting and, oh yes, sensing the weather, may help you bring back some fond thoughts previously misplaced in your memory bank. Whether you reside in the Northern Plains or not, reading this book will leave you with a better understanding of this utterly fascinating, highly dynamic, and in-your-face phenomena known as weather.

Black Elk Peak, Black Hills, SD (South Dakota Department of Tourism)

Dan Lutz

It's difficult to witness weather's impact on our lives, whether by its majestic beauty or its destructive power, without wondering about its beginnings.

🍃 WEATHER'S BEGINNINGS

As surely as the morning sun will rise above the eastern horizon: so too will blinding bone-chilling snowstorms invade the Northern Plains each winter; black ominous storm clouds will reach high over the Plains during late spring afternoons; stifling heat and abundant sunshine will persist through the dog days of summer; and the many colors of autumn will spread a magical and beautiful tapestry across the landscape.

It's difficult to witness weather's impact on our lives, whether by its majestic beauty or its destructive power, without wondering about its beginnings. We contemplate how a puffy cotton ball-like cloud can explode into a dark tornado-producing dynamo in the sky. We question the origin of generous amounts of rain, the genesis of huge bolts of lightning, and straight-line winds with speeds exceeding 100 miles per hour from thunderstorms. And we ponder how the brown landscape of the Northern Great Plain's prairies and canyons transform into various shades of green, red, orange and yellow, year after year.

From where in our universe do these daily weather adventures involving rain, snow, cold, hot, dry and wet, grow and develop? What is it that jump-starts and maintains these weather happenings? In some small way, the origin of each day's weather begins on the wingtip of a butterfly deep in the Amazon rainforest. To a more telling degree, the answer lies in the heavens. Therefore, let me offer up my thoughts on weather's beginning by probing its start in the outer space far away from our lives here on Earth. And while in the heavens, let me convey a few thoughts regarding the sun and moon.

Ol' Sol — Weather's Unmoved Mover

In the lyrics sung by Tommy James and the Shondells, "Let the ball of fire in the sky, keep a watchen over you and I, way up high." Words so very real when it comes to our weather and our very existence here on Earth. There would be no weather and no life, as we know it if that ball of fire in the sky failed to watch over our planet.

The movers and shakers of our weather require energy and lots of it to get the ball rolling. Weather's energy saga begins not anywhere here on Earth or even in the skies above, but far in outer space. It starts within that luminous celestial ball of fire located smack in the middle of our solar system, known as the sun. There, many hydrogen atoms continually slam into one another, forming helium atoms and those all-important byproducts, light, and energy. The light and energy from the sun scatter in all directions, mostly away from Earth, but thankfully a small fraction ends up invading the Earth's atmosphere to fuel the weather engine. The Earth's atmosphere is the envelop of air encasing the Earth and held next to the Earth due largely to Earth's gravitational influence.

For eons, the sun has never quit pumping out this small amount of light and energy into our atmosphere. But here's a thought, "Where on Earth does all this energy end up after its Earth-bound journey through our sky?"

Does yesterday's supply of energy from the sun cease to exist after keeping us warm only to make way for today's fresh energy supply? Or does the sun's non-stop energy supply build up over all those eons in some land body, eventually melting it down to a molten heap? Perhaps it ends up in a massive water body, ultimately boiling it away? Maybe it's used up by the boundless energy demonstrated by children at play?

From the moment the sun first appears over the eastern skyline until its colorful farewell out west, it bombards our piece of the planet with a continuous stream of energy. Not all the sun's energy that started its 93-million-mile journey toward Earth makes it to the surface. The Earth's atmosphere absorbs a full one-fifth of the Earth-bound sun's radiation before it enters the top of that thin layer of gas, known as the troposphere. And it's the troposphere, the lowest 6 to 13 miles of the atmosphere where we live, that confines the weather engine to work its magic.

We live in a world of physical laws. A physical act becomes a law when the same conclusion happens repeatedly based on repeated scientific testing. One such law essential to weather is known as the first law of thermodynamics, which states that energy does not go away. It can be transformed but not destroyed. The energy that traveled 93 million miles from the sun to reach us here on Earth remains with us and not lost. It does not enter stage right and leave stage left. This physical law of thermodynamics tells us that the energy that arrived here on Earth during all the yesterdays will remain with us for all the tomorrows, in one form or another. The second law of thermodynamics states that energy can change its characteristics for the sake of storage. These two laws offer up a clue about where the energy goes after its long journey from the sun and provides the reason why weather happens.

Without confusing everyone with mathematics and diagrams, suffice it to say that when left alone, without outside influences, energy flows spontaneously in one – and only one – direction. Energy flows from being concentrated in one place, to being diffused over a bigger space. Unaided, energy flows from concentrated hot to concentrated cold becoming diffused warm, not the opposite direction. Without the aid of outside forces, energy does not flow from cold to hot.

You witness this taking place when boiling water. The energy/heat from the burner's hot flame moves onto the warm pot, then onto the colder water, diffusing the water to water vapor into a more expansive space. The energy stored in the flame did not disappear but only changed its form to be now stored in the expansive water vapor. Without introducing other outside energy contributing sources, you will never see the now relocated and stored energy in the spread-out water vapor reverse the energy flow back to the original flame's stored and highly concentrated energy/heat.

The energy from the flame was not lost but flowed from being concentrated in the flame, to a more expansive state in the water, then on to the even more expansive and diffused water vapor. By introducing outside forces, such as pressure force, a portion of the energy now stored within the water vapor can be released for other uses.

A locomotive's steam engine is an excellent example of using the energy stored in water vapor to create motion. The locomotive's coal-fired furnace boils the water creating expansive water vapor. By restricting the water vapor's expansion within the locomotive's piston chamber the energy stored in the water vapor transfers to pressure force, driving the pistons to turn the locomotive's wheels, creating motion.

The atmosphere and the Earth take in and stores the sun's abundant supply of energy. Noteworthy storage receptacles include plants, crops, trees, coal, oil and atmospheric humidity or the measure of water-vapor content in the air. It's within these depositories that the eons of the sun's energy are stored waiting to be released.

Nature can be either restless or patient when it comes to releasing this stored-up energy. Nature can wait either seconds, days, months, years, or centuries, before putting the energy to work from its stored-up ordered and concentrated form to the more diffuse and chaotic form. Relative to weather occurrences, the rate-of-release of this stored-up energy dictates either the near-term or the long-term weather happenings.

Kids at play demonstrate an excellent example of a patient nature releasing stored energy. How many times have we asked that universal question, "Where in the world do my kids get all that energy?" Their energy comes from the sun, but not directly. The farmer's maturing corn absorbs some of the sun's energy. The energy, now stored in the harvested corn, is next passed into the cornflakes by the Kellogg people. Finally, the kids get their dose of the sun's energy, needed to play for hours on end, when they consume that bowl of breakfast cornflakes.

Long-term climate change, such as the melting of the ice caps, stands as an excellent example of an even more patient release of stored energy from coal or oil. Ages ago, the sun's energy created the vegetation found in abundance in swamps across the globe. As the plant life died off, new growth took its place. After millions upon millions of years, layer upon layer of vegetation formed. The weight of the soil and water above these layers produced chemical and physical changes to the now highly compressed vegetation, creating coal and oil.

In a very real sense, the sun's energy used to create the vegetation is now stored in the coal and oil. The burning of these fossil fuels then releases a portion of the sun's energy to warm our homes, power our vehicles, and provide electricity. A chemical byproduct from burning these fossil fuels contributes notably to the slow warming of our climate and the ongoing melting of the ice caps.

On the other hand, winter storms, rain/snow, spring thunderstorms, hurricanes, and wind are just a few good examples of an impatient nature's explosive release of stored energy found in the various phases of water. As the sun's energy extracts moisture from the many massive bodies of water found on Earth, via evapotransporation, its energy is transferred into the now created atmospheric water vapor. As water vapor condenses, due to atmospheric pressure changes, into cloud drops, this transition process releases a portion of the stored energy/heat in the water vapor to grow the fair-weather cumulus cloud into a rain shower. Given the right meteorological conditions, a rain shower can grow to heights above the freezing level. Now with the upper portion of the cloud in an environment of freezing temperatures, the raindrops freeze changing into ice pellets. This freezing process releases even more energy/heat stored in the raindrops to the environment intensifying the rain shower even farther becoming the more impressive and powerful thunderstorm.

So, there you have it. The year-round daily supply of sunlight from outer space is the principal energy source necessary to sustain our comfortable lifestyle and to start and maintain the weather engines here on Earth. It is this very release of stored energy, that causes the winter, spring, summer and fall seasons, clouds, rain, thunderstorms, tornadoes, and blizzards; or, we can say it makes the weather happen.

This release of stored energy makes the weather you see, feel, smell, taste, and sometimes hear out your back door. It creates the weather located across your region or country or even around the globe. It makes the weather you experience in the now or the weather you will experience tomorrow, next week, next year or the weather to happen next century.

Without that sudden energy release from water in all its phases aided by outside atmospheric pressure changes, there would be no weather engines directly responsible for the water we drink and the air we breathe, the two essential elements necessary to sustain all forms of life here on Earth. Without old-man sol continuously kick-starting the weather engine here on Earth to do its thing, life as we know it could not exist. So, Mr. Louis Armstrong, that lucky ol' sun has a whole lot to do as it rolls around heaven all day.

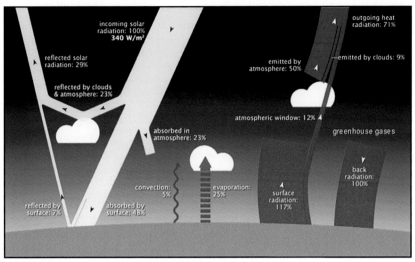

Earth's annual mean energy budget (Google Images)

The Angle of the Sun Causes Wide Temperature Variations

As discussed previously, not all the sun's energy, which started its 93-million-mile journey toward the Earth, makes it to the surface of this planet. The Earth's atmosphere absorbs a full one-fifth of the sun's radiation before it reaches the top of the troposphere, that layer of gas nearest to the Earth and where weather happens. For those packets of energy that do make it into the troposphere, the occupying clouds packed with water droplets and ice crystals reflect much of the energy packets back into space. Air pollution and dust particles found in the troposphere scatter other packets of solar energy, as well. Little of the sun's total radiation that enters the Earth's atmosphere ever reaches its surface, but thankfully, there remains still enough solar energy to operate fully and sustain the weather engine.

The amount of the sun's energy that makes it to the surface varies across the globe. The northern and southern latitudes do not receive the same amount of energy as the mid-latitudes. Since the sun's energy falls upon a round and tilting Earth, and it comes in straight from one point, some regions receive much more energy than needed or wanted. Other areas receive less energy than required but can use more.

The tropics and subtropics located near the equator pretty much face the sun head-on all year long. Therefore, those areas take in more of the sun's energy, and as it turns out, more than needed. The North and South Poles do not face the sun head-on due to their locations on the top and bottom of the globe and, therefore, receive much less energy than needed but can use more.

Without some way of redistributing this ceaseless barrage of incoming solar energy across the globe, the tropics would become much hotter than they are now, while the North and South Poles would become much colder. So, nature resolved this problem by establishing a relatively efficient mode of energy transfer to keep the tropics from overheating and the poles from overcooling.

The daily occurrence of plentiful thunderstorms over the tropics and tropical rain forests does the job of transferring the buildup surface heat energy from near the Earth's surface to the upper levels of the troposphere. Like a pot of boiling water, the strong updrafts found within thunderstorms carry much of the sun's overabundance of energy found in the tropical-heated air to the higher levels of the thunderstorms and into the upper reaches of the troposphere.

From there, the strong mid-latitude upper-level winds rush the heated air, i.e. energy, toward the poles keeping the thermostat turned down in the tropics and turned up at the poles. Vast stretches of water currents found in the oceans also transfer energy away from the tropics. For example, the Gulf Stream, located in the Gulf of Mexico, does an excellent job of transporting tropical energy/heat northward — in this case, northward along the East Coast of the United States.

The continual transfer of heat energy, both vertically by thunderstorms and horizontally by the upper-level winds and ocean currents, keeps the Earth's heat engine in balance, i.e., like Goldilocks's soup — not too hot or too cold. And this continual and dynamic transfer of energy/heat from those hotter tropical regions to the colder polar areas strongly influences the Earth's climate and provides the dynamic environment to maintain its weather engine.

Blemishes on the Sun Produce Breathtaking Displays on Earth

Not only does the sun stir up the weather here on Earth, but those sunspots on the sun's surface stir up geomagnetic storms that trigger the most spectacular light show around — the Northern Lights. Typically, large sunspots measure about two to three times the size of half of the Earth's hemisphere, with a lifetime from a few days to several months. Occasionally, sunspots can reach 14 times larger than the Earth's hemisphere.

Northern lights as viewed over Alaska (Yahoo Images)

It's difficult to believe that the sun has had even more massive sunspots pass across its face. In 1947, those blemishes on the sun's surface registered near 50 times larger than the Earth's hemisphere to become the most massive sunspots ever detected.

There is some debate over who first discovered sunspots. Johann Goldsmid of Holland, Galileo Galilei of Italy, Christopher Scheiner of Germany and Thomas Herriot of England, all of whom claimed to have discerned sunspots sometime around 1611, share the credit. (Dearborn, 1998) All four men observed sunspots through telescopes, but none of them had any clue as to what they were seeing.

Historically, formidable cultural or religiously held beliefs often bring the understanding process to a near standstill. Back then, many stargazers believed the heavens and the sun were pure and blemish-free. They concluded the relatively dark areas on the sun were planets or some other large object passing in front of the sun.

Galileo stepped out of the crowd with a different idea when he exclaimed, "Hmm, there are definitely blemishes on the sun." He noticed the shape of the sunspots changed as they moved toward the edge of the visible sun. He concluded that this would only happen if the spots were on the sun's surface, though he still had no clue just what the objects represented.

After all these centuries, we now know what causes sunspots. A sunspot shows up as a dark spot on the sun's surface because it's an area that remains about 1,600 degrees Fahrenheit colder than its 10,000 degrees Fahrenheit surroundings. Scientists discovered that a stronger magnetic field within the sunspot inhibits the transfer of heat from the much hotter surrounding area into the sunspot itself, keeping the sunspot relatively cooler.

You see, on the sun's surface outside the sunspot area, there exists only gas pressure. And, the higher the gas pressure, the higher the temperature. The interior of the sunspot has a slightly different pressure makeup. There, both gas pressure and a magnetic field pressure field coexist. To keep the pressure balanced between the sunspot and its surroundings, the combination of the magnetic and gas forces within the sunspot keeps the gas pressure lower, and thus a cooler temperature and darker appearance.

Sunspots send out solar flares that eventually reach the Earth's atmosphere. And when these solar flares enter our atmosphere, they can cause all sorts of problems with commercial and military satellite communication links, as well as power grids. Also, for the people who live in the northern latitudes, these same solar flares create dazzling displays of the Northern Lights.

On those few days after peak sunspot activity, sky watchers across the Northern Plains can enjoy nature's most colorful display. On the not-so-frequent days with exceptionally robust sunspot activity, nighttime gazers as far south as Texas, Southern California, and Mexico, too, may be able to view the magnificently bright display of the Northern Lights.

With Weather, Moon a World Apart

Probably the one most inspirational moment this past century occurred on July 20, 1969. On that day the Apollo 11 astronauts Neil Armstrong and Edwin "Buzz" Aldrin (Michael Collins remained in orbit around the moon) stepped onto the surface of the moon. The politics at the time started that journey to the moon, but in the end, the curiosity of the American people sustained it.

Despite Armstrong's "small step for a man, one giant leap for mankind," on this abstruse cold, dry rock, we call the moon, it will always remain a mysterious and stirring celestial body. For centuries, primitive peoples worshiped the moon. Today, it continues to inspire others to create everything from lunar calendars, to love poems and songs, to scary movies about wolf creatures.

Moon-Lite Night to Sun-Lite Day (Don Lutz)

At its closest approach, the moon is 221,460 miles from the Earth, and at its farthest approach, the moon is 252,700 miles from the Earth. Even with these great distances separating the Earth from the moon, an invisible and yet vital linkage exists between the two that affects the well-being of our climate. The ebb and flow of the moon's gravitational forces on the Earth influence the ebb and flow of the ocean tides that, in turn, sway the Earth's climate.

The Apollo 11 mission reported no cheese on the surface of the moon, only rocks and countless numbers of mostly circular impact craters. Depressions caused by debris from passing asteroids, comets, and meteorites bombarding its surface. With only a scant atmosphere on the moon to burn up those moon-bound flying objects from space, they occasionally slam full-force into the moon's rocky surface, creating its pock-marked landscape. Also, with only a negligible atmosphere, there can be no wind, no clouds, nor any precipitation. Consequently, the moon's craters remain unaffected by erosion, leaving them unchanged until another new impact alters the scenery.

If no one is on the moon to hear a meteorite slam into its surface, does it make a noise? The answer to that one is easy — no. No air means no sound, leaving the surface of the moon in complete silence. Also, the moon's sky remains pitch black during the daylight. Without air and wind to kick up dust particles to separate the colors of the sun's light, the daytime sky remains black. Therefore, there can be no blue afternoon sky nor brilliant red sunrises or sunsets. Buzz Aldrin best described the lunar scenery as "magnificent desolation."

The temperature on the moon ranges from light-side-of-the-moon highs of about 265 degrees Fahrenheit to dark-side-of-the-moon lows about a negative 451 degrees Fahrenheit. The moon makes one rotation on its axis in sync with one revolution around the Earth every 27 days, 7 hours, and 43 minutes. This synchronous rotation with the Earth caused by an uneven distribution of mass in the moon (essentially, it is "heavier" on one side than the other) allows the Earth's gravity to keep one side of the moon permanently facing Earth.

The origin of the moon is still a subject of some scientific debate. One accepted theory proclaims that the debris from a massive collision with the young Earth about 4.6 billion years ago formed the moon. A considerable body, perhaps the size of Mars, struck the Earth, throwing out an immense amount of debris that coalesced and cooled in orbit around the Earth.

I, for one, do not think in astronomical terms when gazing at the moon. On those nights when the moon is full, I still look for the old man's face and, on occasion, laugh to myself thinking about the long-running television comedy series, "The Honeymooners," that began with Jackie Gleason's face smiling on the lunar moonscape.

Cycling on the Moon

Despite the on-going precise celestial merry-go-round of the Earth circling the sun in 365.2421891 days and the moon orbiting the Earth in 27.322 days, the Earth and moon will never return to their exact starting position — relative to the stars. Our entire solar system, as a whole, travels completely around the Milky Way galaxy in about 230 million years. Therefore, on any given day falling within this period, neither the sun, moon, nor the Earth will ever return to the same spot – relative to the stars. Even after the solar system's 230-million-year trip around the Milky Way, an expanding universe dictates that any member of our solar system will never return to the same place, ever.

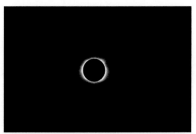

Total Solar Eclipse, August 21, 2017 (Orville Hengen Jr.)

Staying within our solar system, the Earth and moon future locations relative to the sun, occur in a timely and predictable fashion regardless of their whereabouts within the Milky Way. This allows all of us star gazers to plan, observe and experience both the lunar and solar eclipse, firsthand.

A solar eclipse happens when the moon passes in front of the sun, bringing a nighttime feel to the Earth for a brief time. On August 21, 2017, a total solar eclipse, lasting approximately 2 minutes and 40 seconds at any one location, traversed this entire country. Fourteen states enjoyed the experience of a total solar eclipse on that day. I observed the August 2017 total solar eclipse at the Agate Fossil Beds National Monument, located in Western Nebraska. The last solar eclipse to crisscross the entire U.S. happened a century ago on June 8, 1918. The exact location and time of the next solar eclipse to occur in the U.S. is calculated to happen on April 8, 2024.

A lunar eclipse occurs when the Earth passes between the sun and moon, casting its shadow across the surface of the moon. The soon-to-be-eclipsed moon starts as a big bright sun-lit full moon in the night sky, and over a short period, takes on the presence of a three-quarter moon, then a quasi-half-moon, followed by a quarter moon and finally, during the total eclipse, the appearance of a new moon. Even though the lunar eclipse phases of the moon appear like the more familiar monthly lunar phases, they both are the result of different heavenly alignments.

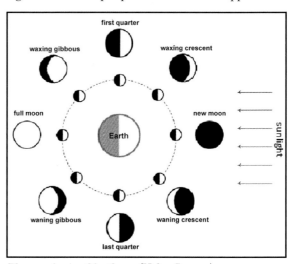

A monthly new moon, albeit present high in the night sky, is nearly invisible. It's not the Earth's shadow hiding the moon, but the moon's orbit located in such a position in the night sky, where from our viewpoint on Earth the sun is

The moon's monthly phases (Yahoo Images)

unable to light it up wholly or partially. The darkness of the new moon has nothing to do with the Earth's shadow covering its surface, nor the dark "far side" of the moon somehow rotating around to face the Earth. It has everything to do with how the sun illuminates its moon's surface that faces the Earth.

As the moon circles the Earth from one night to the next, its night sky shape appears to change as the sun progressively lights up more and more (or less and less) of the moon's surface. The moon's monthly shape changes from a new moon, to a crescent moon, to a half moon, to a gibbous moon, to a full moon and back again.

As mentioned previously, it takes 27 days, 7 hours, and 43 minutes to complete the moon's monthly cycle. Since the lunar cycle time frame does not coincide precisely with the number of days found in calendar months, on occasion two full moons can occur during a single month. When this happens, the second full moon is called a "blue moon." Thus, the saying, "Once in a blue moon."

Over the years, the full moon has acquired different names depending on the time of year it appears. For instance, the "harvest moon" is the full moon that appears nearest to the autumnal equinox in late September or early October. The harvest moon, in conjunction with the sun, can provide the farmer with the year's maximum amount of light over 24 hours. You see, the harvest moon rises over the eastern horizon just about the same time the sun sets behind the western skyline, allowing the farmer to enjoy extended light, albeit lunar light, to harvest well into the evening hours.

Some other full moon names are:

"**January** moon after yule or old moon."
"**February** snow moon, hunger moon, or wolf moon."
"**March** sap moon, crow moon or Lenten moon."
"**April** grass moon or egg moon."
"**May** milk moon or planting moon."
"**June** rose moon, flower moon, or strawberry moon."
"**July** thunder moon or hay moon."
"**August** grain moon or green corn moon."
"**September** fruit moon or harvest moon."
"**October** harvest moon or hunter's moon."
"**November** hunter's moon, frosty moon, or beaver moon."
"**December** moon before Yule, or long night moon."

A Once in a Lifetime Moon Glow

The winter solstice officially starts the Northern Hemisphere's coldest season. It's the day of the year with the least daylight and the most darkness. Tuesday, December 22, 1999, however, proved to be different when winter's first full day of illumination lasted a little longer than usual, ever so slightly pushing back the nighttime darkness.

Unlike so many previous and following winters' solstice, a full moon illuminated the skies high above on December 22, 1999. And not just any full moon, but a vast and beaming full moon. The celestial geometry of the heavens on that day, and perhaps a little Y2K thrown in for grins (I'm only kidding about the Y2K), significantly contributed to the size and brilliance of that moon.

The unusually large image of the moon occurred on that date for a couple of reasons. The lunar perigee, or the point in the moon's orbit where the moon is closest to Earth, coincided with the winter solstice. Therefore, the moon appeared about 14 percent larger that night than it would at apogee or the point in its elliptical orbit farthest from the Earth. And with the Earth several million miles closer to the sun during winter than summer, 7 percent additional sunlight illuminated the moon's surface, making it even brighter.

Given a clear and snow-covered night that evening across Western South Dakota, the exceptionally bright lunar light alone could possibly suppress the need for streetlights. Folklore tells us that on December 21, 1866, the Lakota Sioux took advantage of this combination of lunar and solar configurations to stage a devastating retaliatory ambush on soldiers in the Wyoming Territory. Unfortunately for those soldiers, this infrequent lunar event help to contribute substantially to their demise. And too, the December 22, 1999, evening's super bright full moon, will not return for another 100 or so years.

Gravity: Are We in Grave Danger?

Adults appreciate the power of gravity, but kids could and do care less. I think it's because kid's bodies grow upwards, defying gravity. Adult body parts will eventually succumb to gravity's downward forces, albeit with much rebellion. Even though many of my body parts stopped reaching for the stars some time ago, gravity continues to redirect a few portions downward.

Gravity not only keeps us glued to the Earth but, according to Albert Einstein, it warps and curves the universe keeping the moon orbiting the Earth and the planets orbiting the sun. Gravity gives meaning to a lot of what we see happening around us here on Earth and high in the heavens.

Newton's falling apple (Yahoo Images)

Sir Isaac Newton sighting of the falling apple while relaxing under a tree, started the gravity craze. While contemplating the reason explaining why planets have elliptical orbits and not round orbits and thinking back about that falling apple, Sir Isaac Newton put forth a rational theory regarding gravitational force. He concluded that all bodies attract each other, with the larger bodies having the strongest attraction. Also, this mutual attraction among bodies decreases with distance.

It now stands to Newtonian reason that the mutual attraction between the falling apple and Earth moved the Earth toward the apple and the apple toward the Earth. With the Earth so much larger, and with so much more mass than the apple, the Earth moved only an imperceptible distance, but the apple did in fact move the Earth.

The simple act of describing why an apple would fall to the Earth below had an enormous impact on how we began to view the heavens above. This one conundrum stands out from the rest: If all objects are attracted to one another, then the universe should collapse onto itself.

Undoubtedly, the gravitational pull of more massive suns pulling on other less massive suns would eventually cause them to collide, forming an even more massive sun with an even stronger gravitational force, pulling an ever-higher number of suns into its growing mass. Soon all suns would be confined into one extremely dense mass signaling the end of the universe.

Thank goodness there's no need for panic. The universe does not appear to be collapsing onto itself. But why not? Some argue for an infinite universe, which means an endless number of suns tug equally from all directions on other suns, canceling their influence on any one sun.

Not so fast, says the German philosopher Heinrich Wilhelm Matthias Olbers. He argued that the universe cannot be infinite because the rays of light — no matter how dim — from an endless number of suns would light up the night sky. Okay, good point, but the dark-night-sky believers counter that by saying there must be some absorption material blocking the rays from the distant infinite suns. Also, a good point, but not likely because any absorption material would eventually heat up, finally giving off light as bright as any sun and again lighting up the night sky.

In 1929, Edwin Hubble explained away this falling-apple, universe-ending, gravity dilemma. After years of study and careful observation, he observed distant galaxies moving away from one another at tremendous speeds and not attracted to one another. Hubble's study implies an expanding universe created by the Big Bang, which defies gravity's pull.

Picture in your mind, if you will, spots on a balloon representing galaxies and the balloon depicting our universe. Now, as the balloon expands, the dots move farther apart. An expanding universe also can explain away the night sky. Those light rays from distant suns that would have streaked toward our darken night sky to brighten it up, now either retreat or remain in place as their sun's source races away in one direction while Earth moves away in another direction.

It's possible that at a point in time in the distant future, the stars that we enjoy gazing at during the evening will disappear as they speed away from our Earth-bound vantage point. And, at that time far into the future, sadly it would appear to us that we are alone in our universe.

Dan Lutz

Just as these waves of water spread out across the pond, so do pressure waves of weather spread across the atmosphere. Most importantly, we need to appreciate the sea of gas surrounding the planet, called the atmosphere, within which these meteorological pressure undulations can pass.

🍃 THE ATMOSPHERE

A stone tossed into a pond sends out perfect concentric expanding waves on the surface of the water only to disappear with time and distance. Just as these waves of water spread out across the pond, so do pressure waves of weather spread across the atmosphere. Like the stone-produced waves of water on the pond, the weather is first kick-started, then grows, matures, propagates and fades away, on those waves of pressure traveling through the atmosphere.

Pressure waves of the weather traveling through the atmosphere cover a spectrum of scales from periodic puffs of wind drifting across the prairie to global dimensions that describe the climates worldwide. To fully understand these waves of weather, we first must realize that our perspective of weather's actions is from a spinning nomadic planet traveling around the sun. And secondly and most importantly, we need to appreciate the sea of gas surrounding it, called the atmosphere, within which these meteorological pressure undulations can pass.

Weather Occurs in a Sphere of Gas

The gaseous envelope surrounding our Earth, known as the atmosphere, consists of four reasonably distinct layers. Each layer gets one of its characteristics by how temperature within each layer changes with elevation.

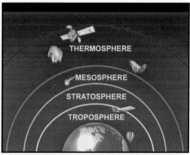

The layer closest to the Earth's surface — and the shell where we enjoy picnics, softball games, and family gatherings, is referred to as the troposphere. Its thickness is between 6 to 13 miles.

The different layers of the Earth's atmosphere (Yahoo Images)

Throughout the troposphere, the ambient air temperature typically decreases with height. Temperatures are normally warmer in the lower part of the troposphere, near the Earth's surface than at higher elevations.

The Earth's weather happens within the six- to thirteen-mile thick troposphere arena. Cloud types from fair-weather cumulus clouds, to ice crystal-packed cirrus clouds, to the massive cumulonimbus thunderstorm clouds, reside in this relatively thin and fragile layer of gas closest to the Earth. Gravity's pull traps about three-fourths of the atmosphere's total weight (yes, the atmosphere has weight) and practically all of its water content within the troposphere where we live out our lives.

At the upper limit of the troposphere, one finds the tropopause, a stable ribbon of air separating the troposphere, from the next higher layer of the atmosphere, known as the stratosphere. The stable tropopause marks the sudden change in rate at which temperature drops with height., causing the rising air to cease cooling with height. Therefore, the tropopause acts as a ceiling to "cap off" growing thunderstorms, producing the characteristic flat wispy-looking anvil top blowing off and away from mature thunderstorms.

In those areas of severe weather, the time-lapse satellite sequence of developing thunderstorms clearly shows the tallest growing cloud tops reaching the tropopause, some nine miles up or higher, only to spread out flat and drift downwind. Think of a thunderstorm as a glob of silly putty thrown up to the ceiling of your room. The putty will flatten out when it hits the ceiling, as does the large cumulonimbus cloud formed by a vertically growing mature thunderstorm flattens out when it hits the stable tropopause.

The next layer of gas above the troposphere and tropopause and below the mesosphere is the stratosphere, located from 5 to 7 miles up to 15 to 18 miles above the Earth's surface. This second discrete layer of the atmosphere exhibits little if any change in temperature with increasing height. Of note, a high concentration of ozone at the top of the stratosphere, which serves as a highly efficient absorber of the dangerous ultraviolet rays from the sun, keeps that part of the stratosphere considerably warmer than the rest of the lower layers.

In 1934 and 1935, the Army Air Corps and National Geographic Society launched manned balloon flights into the stratosphere to a record 72,395 feet, nearly 14 miles. The launches originated from the Spring Creek valley region known as the "Stratobowl" located in the Black Hills west of Rapid City. These balloon-bearing gondola research flights into the stratosphere are considered the beginning of the space age.

November 11, 1935 balloon flight to the stratosphere launched from the Stratobowl, Black Hills, South Dakota (H. Lee Wells, Jr.)

On top of the stratosphere is the mesosphere, ranging in altitude between 12 and 50 miles. Temperatures in the mesosphere first increase with height then decrease. At the upper limit of the mesosphere, or about 50 miles up, most meteors burn and disintegrate as they try to penetrate our atmosphere from outer space.

The outer most layer of the atmosphere is the thermosphere, reaching way beyond 300 miles from the surface of the Earth. A part of the thermosphere extending from 50 to 300 miles high is the ionosphere. Here particles that make up the ionosphere can and do reflect radio waves, which on some nights allows us to hear radio stations transmitting from thousands of miles away.

Even though changes in temperature and density from all these layers can impact our weather near the surface, we focus our weather and forecasting energies on the troposphere, the envelop of gas where weather happens. And because we make our nests to live out our lives in this relatively thin and delicate blanket of air, we need to remain vigil to keep it unpolluted for future weather watchers.

It's Cold Up There, Icarus

With the top of the troposphere closer to the sun than the bottom, one would believe it to be warmer than the bottom of the troposphere. Recall when Icarus fled from Crete to Sicily using wings made of wax and feathers. Unfortunately, his flight to the top of the troposphere took him much too close to the sun, heating then melting the wax, and the adventurous youth plunged to his death. This parable, albeit enchanting, diverges from experience and fact.

Icarus Parable (Yahoo Images)

A hot air balloonist knows that the higher one flies in the troposphere, the colder the temperature, not warmer. And most of us who have lived in mountainous regions long enough know that on average, the temperature on top of the mountain measures more often cooler than the same-day temperature at the base of the same mountain. So why colder mountain peak ambient air than the mountain base ambient air? And how can the tallest mountains with their peaks closest to the sun remain snow-covered year-round, though snowmelt takes place lower down farther from the sun?

So, what gives? Well, let me tell you air gives like flexible spongy putty. It can be compressed (causing the air to warm) or stretched (causing the air to cool). Like that glob of flexible silly putty stuck flat on your ceiling, the air is pliable. So, because air can stretch or expand, air temperatures cool with height.

Here's how it works: The rays of the sun pass through the transparent atmosphere striking and heating the Earth. In turn, the ground warms the air just above its surface. Therefore, like a pot of boiling water on the stove, the atmosphere, for the most part, is heated from below.

But wait a minute, after heating a pot of water from below, the heated water moves to the top and replaced at the bottom by the relatively cooler water from above. Given enough time, the entire container of water develops a uniform temperature from bottom to top. If the pot becomes evenly heated, why not the atmosphere?

The bottom heating in the pot brings into play the heat transfer method known as convection. Convection is motion within the water that results in the transport and mixing of certain properties of the water. Convection does the job of redistributing heat across the container of water, with time, leaving the top of the pot heated to the same degree as the bottom.

The air in our atmosphere becomes heated like the water in that pot. The sun heats the Earth, which in turn heats the air from below. The Earth-heated air parcels near the surface rise, due to convection, bringing the heat to the higher elevations. So, one would think that after a few hours of active afternoon heating and mixing, the warmth near the surface of the Earth would be well-mixed throughout the atmosphere, warming the mountain peaks.

Not so. Let's get back to that glob of silly putty, i.e., flexible air that can stretch or expand. Unlike the heated water in the pot, a parcel of air riding the convection currents upward will expand, because the farther up it goes it finds itself with less weighted air pushing against it. So in uncomplicated terms, with less and less air pushing against this rising parcel, it uses up some of the energy retrieved from the heated Earth to expand outward and this loss of energy due to the expansion cools the air parcel.

As convection lifts these expanding air parcels high into the troposphere to levels of less pressure, the air temperature lapses or, in other words, becomes cooler with height. Like Icarus falling from the skies because he lost the wax to hold his wings together, the temperature of the rising air parcel will tumble as well. So, when visiting high places, like Black Elk Peak in the Black Hills or Pikes Peak in the Rocky Mountains, expect a slight chill, not only from the view but from the surrounding air.

The Air's Gases and Role They Play

Two gases make up 99 percent of the dry, odorless and invisible air comprising our atmosphere: Nitrogen (79 percent), and Oxygen (20 percent). Much smaller quantities of Argon, Carbon Dioxide, Water Vapor, and Ozone gases also can be found in the air we breathe.

Of these, oxygen sustains the biological functions of the animal world, both large and small. Plant life across the globe depends on carbon dioxide.

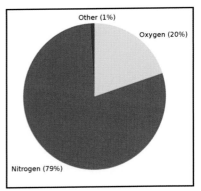

Breakdown of the air's gases found in the Earth's atmosphere

Finally, an odorless and invisible gas, whose importance far outweighs its small percentage of the lower atmospheric gaseous makeup, is water vapor.

No water vapor means no clouds floating across the sky and without clouds, no subsequent precipitation development. No precipitation means no water, and no water means no life.

Water vapor also provides the nighttime blanket that traps the day's solar heating and keeps a portion of its warmth from escaping back to outer space. Carbon dioxide also contributes to the preservation of relatively warm evenings and the lowering of our nighttime heating bills.

Another minor gas ingredient of our atmosphere with a significant implication in the recipe of life-sustaining gases is the photochemical by-product of oxygen known as ozone. Ozone is a colorless gaseous form of oxygen containing three atoms in its molecular makeup instead of the two found in oxygen.

A thin layer of ozone lies in the upper atmosphere around 15 miles above the Earth. There, the ozone regulates the types and amount of solar radiation that reach the Earth's surface, and in doing so, filters out much of the sun's dangerous ultraviolet radiation. Too much ultraviolet radiation has the potential to increase the incidence of skin cancer and cataracts in humans, harm crops, erode man-made century-old structures, and interfere with marine life. The acute properties of oxygen and its interplay with the sun to form ozone high up in our atmosphere makes our lives on this planet possible.

In 1999, measurements of the ozone layer using the latest satellite instruments discovered an "ozone hole," a region of ozone depletion over Antarctica, about the size of North America. To combat this apparent growing ozone hole, the United Nations has in place measures to reduce the amount of ozone-depleting chemicals released into our atmosphere. Scientists expect these measures to produce a period of slow recovery, with full recovery predicted by about the year 2050. Latest measurements confirm the ozone hole is shrinking in size.

The Strange and Wonderful Properties of Water

Everyday water persists as the most vital and yet most bizarre substance found in our universe. When astronomers look for life as we know it on other planets, the first question they ask themselves, "Is there water?" Here on Earth, water surrounds us, in the atmosphere, on the surface, and beneath the surface. The most abundant substance found in our bodies is water. Because water contributes so much to our being, it genuinely and troublingly affects our psyche.

I say this because I have observed rational people shop the supermarket aisles for water. As much as I enjoy maximizing my food value, no coupon nor any "buy one get one free" special could convince me to purchase odorless and tasteless water off the shelf. How in good conscience can people complain about the high price of gas, all the while sipping $6-per-gallon of water? The genuinely perplexing fact about all this is that water fountains gushing with free water can be found in the back of most supermarkets.

Besides engaging our psyche, water also exhibits extraordinary powers over the weather. Nature uses water in all its forms, i.e., gas, liquid and solid, to store a portion of the sun's energy to initiate later the beginning of many weather-making processes. The water vapor found in the troposphere stores the energy needed to jump-start the weather engine.

To clarify the importance of water in the atmosphere, a calorie is the amount of energy/heat it takes to raise the temperature of one gram of pure water one degree Celsius at sea level. Therefore, at sea level it takes 100 calories to heat one gram of water from zero degrees Celsius, the melting point of ice, to 100 degrees Celsius, the boiling point of water. To transform one gram of water into a gas, i.e., water vapor, requires 540 additional calories of energy/heat. So, with copious amounts of water, in all its phases (gas, liquid and ice), found in the troposphere, abundant amounts of the sun's energy can be stored and made available to make weather. Let me explain further.

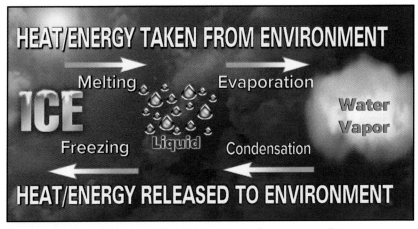

Heat/Energy transfer between the environment and ice, water and water vapor

Much of the inflow of the energy/heat needed to change atmospheric ice to water and atmospheric water to gas comes from the sun. Because the first law of thermodynamics states that energy is not lost, nature stores the sun's energy in the abundant water and the water vapor found in our atmosphere. On the return trip by utilizing outside forces such as pressure or changes in ambient air temperature, the atmosphere can transform the water vapor back to the water, or the water back to the ice, or even the water vapor back to ice, skipping over the water phase. All of these reversals release the stored sun's energy/heat back to the environment to run the weather engine.

In simpler terms, converting ice to liquid water (the melting process), and liquid water to water vapor (the evaporation process) requires a lot of energy/heat taken from the environment – cooling the environment. Returning water vapor to liquid water (the condensation process) and liquid water to ice (the freezing process) gives off a lot of energy/heat back to the environment – warming the environment. This back and forth exchange of energy/heat through the various water phases (gas, liquid and solid) gives the troposphere, the atmospheric layer of gas within which we live, the energy to create afternoon clouds, evening showers, snow, hail, and brisk winds from a distant thunderstorm.

Ever wonder why on a chilly early Spring morning with air temperatures hovering slightly warmer than freezing and a light rain falling, that on some occasions the light rain changes into light snow right before your eyes, without any measurable penetration of colder air. The answer rests with the evaporation of the raindrops. Given the right atmospheric conditions, the evaporation of the raindrops removes energy/heat from the surrounding air, thus cooling the air to a temperature cold enough to allow the remaining water drops to transition into sleet or snow. On occasion, as morning becomes afternoon the day's warming of the surrounding ambient air may reverse the process, allowing the light snow to return to light afternoon rain.

Unlike most other liquid substances, water expands when it becomes solid. For nearly every other known chemical compound, molecules are held closer together in the solid state than in the liquid state. Water, however, is unique in that it bonds in such a way that the molecules expand farther apart in the solid form (ice) than in the liquid. (Saffer)

For this reason, ice floats on the water allowing fish to survive the winter. Because of ice forming on top of the lake, it effectively keeps the rest of the lake below from freezing solid, trapping the fish into an icy tomb. If the ponds, lakes or oceans ever fill with ice, life there would not be possible.

Water also gives us the brilliance of the seasons. The colors of the seasons are born from this colorless substance known as water. Deserts come alive with color shortly after a spring rain. The spring showers work like an artist's paintbrush, transforming our South Dakota brown prairies into a lush dark green splashed with a tapestry of wildflowers giving off vivid colors such as reds, yellows, blues and purples across the vast grasslands. And, of course, there would be no colors of the rainbow without water.

Hey, perhaps that $6 per gallon water isn't such a bad deal after all.

Air in Atmosphere Behaves Like Water in a Creek

Often during the summer season, a few of my mornings are spent fly fishing in Rapid Creek with my dog, Wally. My fishing excursions go like this: I do the fishing with my trusted fly rod, shorts, and waterproof sandals, while Wally splashes up rocks from the bottom of the creek. Wally and I, me with the rod in hand and Wally without, wade into the creek to pass a part of our mornings in search of the "elusive trout."

If you fly fish, you understand and appreciate the expectation of catching that enigmatic trout after spending hours standing in a cold creek. To hook, land and release that one elusive trout requires careful study and proper reading the water's surface to identify the most likely trout location in the creek. It also involves determining the appropriate fly that best simulates the day's bugs and finally, making the proper fly presentation to the trout. A lot to think about, but I believe it's the reason fly-fishing-people fly fish.

A View of Rapid Creek in South Dakota

Spending a couple of hours alone, standing not quite knee-deep in Rapid Creek, allows my imagination to drift easily into the realm of the abnormal. While watching the water pass just below my kneecaps, I cannot help but think about the similarities between the undulating water in the creek and the undulating air in our troposphere. The same weather patterns found on typical weather maps, such as fronts, highs, lows, and wind patterns, also happen in the water flowing by me in the creek.

Despite water being a liquid and air a gas, they both abide by the universal laws of fluid mechanics and flow dynamics. The big difference between the two is the property called density, which is the ratio of the mass of any substance to the volume it occupies. It's more difficult to walk against moving water in the creek than moving air because water has a much higher density, or, in other words, water molecules are more jam-packed compared to air molecules.

To illustrate the similarities of the atmospheric air and creek water, a deep pool of water, home to many trout, compares well with a large dome of high pressure in the atmosphere. The deeper the dome of air in the atmosphere, the higher the pressure near the surface. Similarly, the deeper the water hole, the higher the pressure at the creek's bottom. Any fisherman will tell you the current slows in the middle of a deep pool and, likewise, the air moves slowly in the center of a large dome of high pressure.

However, when the dome of piled-up air squeezes against an area of low pressure, the wind increases just as the current increases around that deep blue waterhole. The relatively shallow and faster water piles up and slows when moving into the deep waterhole and increases in speed when flowing out of the deep waterhole into the shallower depths (less water pressure) found along the waterhole's down-stream edges.

While standing in Rapid Creek, I could not help but notice the turbulence behind large rocks, where trout often station themselves to conserve energy against the fast-moving water. Similar atmospheric properties occur across our region due to the Black Hills. When the fast-moving air strikes the Black Hills, it either goes around, over the top or both, often causing turbulence on the downwind side of the Black Hills. These areas of disturbance, like the ripples found behind the rocks in the creek, usually foster bands of wave-like stationary clouds, turbulent standing lenticular clouds, and possible shower activity in the downwind sky.

Mountain-induced standing lenticular wave cloud, Black Hills, SD

Mountain induced wave-like stationary clouds (Google Images)

Rain or snow showers are the one atmospheric characteristic not found in Rapid Creek. Even so, trout can forecast rain showers because, from my experience, they tend to attack the fly more often just before a storm. I don't like to fish before an approaching storm. Maybe it's because I do not want to get the upper half of my body wet, or perhaps it's because I do not want to improve my chances of catching that "elusive trout." You see, if I ever hook that one wise trout hiding in the perfect spot in the creek, I just might lose the reason to fly fish. I'm only kidding.

The Waters Define the Equator

The forces that move and shake our weather include pressure, gravitation, friction, and Coriolis force.

To better illustrate these forces in action, picture Tiger Woods about 180 yards from the green. The pressure force from striking the golf ball with his eight iron propels the golf ball down the fairway toward the green.

The Coriolis force will cause even a Tiger Wood's golf ball to drift to the right, or fade, in the Northern Hemisphere and move left, or draw, in the Southern Hemisphere, albeit slightly, as it soars through the air. The force of gravity pulls the ball down from its flight, allowing the force of friction to help stop the ball inches from the cup. In weather, like in golf, the pressure moves, Coriolis turns, gravity pulls, and friction stops.

These same forces that affect golf balls, also impact air movement in the atmosphere and water movement in the oceans. The Coriolis force only exists because the Earth spins, and this rotation action turns the winds clockwise around a high-pressure system and counterclockwise around a low-pressure system in the Northern Hemisphere. The reverse holds true in the Southern Hemisphere. Keep in mind that the Coriolis force moves air and water to the right in the Northern Hemisphere and to the left in the Southern Hemisphere.

The Coriolis force can be observed almost everywhere on Earth except at the equator, where it has no impact whatsoever. Pressure, gravity, and friction can all do their thing, pushing and pulling air masses and water bodies about the equator, but not the Coriolis force. Only at the equator can Tiger Woods hit a golf ball truly straight down the middle of an east or west facing fairway.

Though Coriolis force is absent at the equator, it plays a vital role in making that imaginary equatorial line circumventing the store-bought globe visible in the real world. Considering the force of Coriolis on the waters a few degrees latitude north and south of the equator, it nicely outlines the location of the equator by both a robust narrow subsurface current and ribbons of cold surface water visible from infrared satellite photos.

Here's a little-known actuality, the ocean is not level. The Western region of the Pacific Ocean is higher than the Eastern area. As the more upper Western region of the Pacific waters flows downhill toward the lower Eastern area of the Pacific, the Coriolis force turns the water to the right, preventing it from flowing directly downhill from west to east due to gravity alone. However, along the equator, due to the absence of the Coriolis force, the water does indeed flow straight downhill from west to east, in an intense, narrow band confined to the immediate neighborhood of the equator.

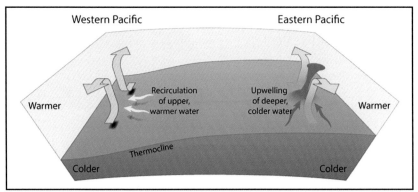

Parting of the trade winds along the equator (Yahoo Images)

Because the northeasterly trade winds (winds blowing from east to west) in the Northern Hemisphere drive the sun-warmed surface water at the equator back to the west, literally uphill, the subsurface gravity-driven downhill current at the equator ends up colder at about 100 meters, unaffected by the trade winds. Also, the southeasterly trade winds, found in the Southern Hemisphere, also drive the surface water at the equator to the west.

The Coriolis force just north of the equator deflects the westward-bound surface water to slice right away from the equator. At the same time, the Coriolis force deflects westward-bound surface water just south of the equator, now in the Southern Hemisphere, to slice left away from the equator.

This parting of the equatorial surface waters brings about an upwelling of cooler subsurface water directly along the equator, making the exact location of the equator visible by temperature-sensing infrared satellites.

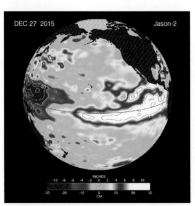

Infrared satellite image showing upwelling along the equator (Yahoo Images)

The Doldrums, No Place for Man or Beast

Samuel Taylor Coleridge penned on a piece of paper, "The Rime of the Ancient Mariner," which contains the following words, "Water, water, everywhere, and all the boards did shrink; Water, water everywhere, nor a drop to drink."

We do not know where this unfortunate mariner got stuck drifting aimlessly on the ocean blue, but a good guess would be somewhere between the zones of the westerly trade winds and easterly trade winds. In between those two zones rests the dreaded calm that caused many a sailing vessel to become adrift for many days on end. Commonly found along the 30-degree latitude line, this area of dead air became known as the "horse latitudes."

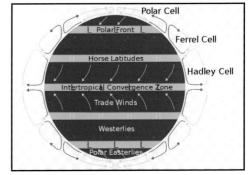

These latitudes were dubbed the "horse latitudes" because desperate sailors trying to sail from the Old Spain to the New World would throw overboard their cargo of horses in a frantic attempt to pull their stranded ship back into the trade winds in a futile effort to continue their journey.

The Horse Latitudes located on and about the 30-degree north latitude line (Yahoo Images)

Those sailors would set sail from the south of Old Spain to take advantage of the easterly trade winds found in the tropics and return to Spain from the New World on a more northerly route to catch westerly trade winds found in the middle latitudes. To navigate from one trade wind zone to the other, those knowledgeable sailors had to steer their vessel through the dreaded calm, hot and dry horse latitudes.

Sailors learned about these areas of dead air tucked away between the east and west trade winds from other sailors' tales of woe or from the realities of their own harsh experience. They knew the "horse latitudes" existed, but not why the dead calm until around 1735, when a London lawyer, George Hadley, proposed an explanation that proved to be mostly correct.

Hadley stated that because of a hot equator, the warm, light air rises into the upper levels, and heads poleward where it cools and sinks over the cold North Pole. Well, as Hadley pointed out, the equator-launched air never really makes it to the North Pole directly.

Hadley understood that the rising hot air over the equator cools due to expansion as the air pressure decreases. He reasoned that after a period, the air traveling upward and northward, in fact, cooled to a point where it sank not at the North Pole, but along the 30-degree latitude. The sinking air strikes the ground (or water in this case) and spreads in all directions. The northward spreading air turns to the right, due to Coriolis force ensuing from the Earth's west to east spin, and thus setting up the westerly Trade winds (winds blowing from west to east) in the mid-latitudes. The southward spreading air also turns to the right, again due to Coriolis force, but now the air turns toward the west, setting up the tropical easterly trade winds (winds blowing from the east to the west).

As the air sinks along the 30-degree latitude, whether over land or sea, it heats and dries due to compression, leaving little chance for cloud development, no less the possibility of rain. The sinking air traveling from lower air pressure aloft to higher air pressure near the surface, not only increases the air pressure creating warm temperatures and dry conditions, but also clear blue sunny skies and little, if any, wind.

For this very reason, many of the Northern Hemisphere deserts are located along the 30-degree latitude. Even if the Ancient Mariners were not at sea, but landlocked along this latitude, they would still have discovered, "Nor a drop to drink."

Nighttime Luminous Clouds Light Up Sky

Very high in the western sky on June 22, 1999, another of nature's spectacular light displays occurred to the delight of those onlookers fortunate enough to have witnessed it. Many people who lived in the Black Hills region for years may have noticed the light show before without full appreciation for what they saw across the darkened horizon.

Noctilucent clouds (Google Images)

That glorious light shows of noctilucent clouds spread across the night sky starting at 9:45 PM and persisted until 11:00 PM when the ghostly illumination slipped away below the horizon.

These occasional sightings of noctilucent clouds at the Black Hill's latitude easily match, in beauty and colorful display, those more common breathtaking Northern Light displays found at higher latitudes. The two Latin terms "nocti" and "lucent" loosely translate to night and luminous, or simply clouds that shine at night.

These beautiful and mysterious clouds ride high above the highest clouds associated with weather, well above 99.9 percent of our atmosphere. The more common weather clouds confine themselves to the 6- to 13-mile-deep troposphere, or the relative thin envelop of gases closest to the surface of the Earth. Noctilucent clouds drift along at the very top of the mesosphere, at an altitude approximately 40 miles above the troposphere.

These noctilucent clouds skirt the lowest fringes of the aurora and yet remain above the height where meteors typically burn up before striking the Earth. For reasons not fully understood, these clouds only display their wispy light show at the higher latitudes, along the 50- to 65-degree latitude range, north or south of the equator, and during the summer months, peaking during June and July over the Northern Hemisphere.

With the sun seven to 10 degrees below the horizon, only these clouds, highest of all clouds above the Earth, remain illuminated by the sun. Against the black night sky, noctilucent clouds appear wave-like with a wispy blue to white color that changes quickly with time. These clouds are so thin that starlight shines through with only a slight decrease in brightness.

Scientists' best guess regarding these clouds' makeup paints a not-so-colorful picture of meteor dust particles covered with ice. It's the sun's bending of light by these particles that cause the clouds to appear luminous.

Many questions about this rare and beautiful spectacle remain unanswered. It's OK in my mind that some divine mysteries remain unknown. So, on those nights when these luminous clouds appear, why ask why? Let's simply enjoy one of nature's magnificent shows.

Badlands Upside-Down

On November 30, 1999, a Fata Morgana sighting occurred stretching above the eastern horizon over the Badlands.

In my many decades studying and working in the field of meteorology, I've observed a lot of strange looking atmospheric phenomena, but never a Fata Morgana. Unless you love trivia with a strong passion for appearing on the next "Who Wants To Be A Millionaire," assuredly you ask yourself: "What in the name of the cosmos is a Fata Morgana?" And "Why was this manifestation levitating so darn close to the Black Hills?"

A Fata Morgana is not some UFO adversary looming on the horizon, but an exaggeration of often everyday objects, like cities, bodies of water and mountains. The legend goes like this: The notorious Morgan LaFee, enchantress and sister of King Arthur, created misleading images of large cities to baffle the enemy's armadas. These hallucinatory cities of abundance and amazement looked genuine from a distance, only to dissolve quickly under closer scrutiny.

Specific atmospheric circumstances must be in place to form these deceptive, larger than life, and misleading images referred to as Fata Morgana. This set of weather conditions happens most often in the Strait of Messina, Italy, and over the Great Lakes. However, back in 1999, on that November day, the Fata Morgana made its way to West River, making the Badlands' landscape even more impressive and majestic.

Recall the ballad about Sam and that big green tree where the water's running free, waiting for you and me…water…cool, clear water. Well regrettably, old Sam had his first and possibly his last lesson about false images. To Sam's surprise and ultimate dismay, his vision turned out to be imaginary. The Badlands' Fata Morgana image, however, unlike Sam's, is a real vision that appears to be in a different location that just so happens to be inverted.

It's possible that Sam's extreme thirst, coupled with mental and physical strain, gave rise to this psychological aberration of cool and clear water. Nevertheless, a specific set of atmospheric conditions, and not the onlooker's imaginings, caused the Badlands to appear out of place, inverted, and more prominent than normal.

Fata Morgana image (Google Images)

On this day, the winds blowing westward off the prairies toward the Black Hills established a cool layer of air at the surface, displacing a warmer air layer near the surface upward. The boundary between the two layers of air with different densities produced an atmospheric lens-like effect, bending, and magnifying the rays of light from the distant Badlands. Similar to the way the lens of a pin-hole camera inverts an image on the camera's back wall.

In the bizarre image, the Badlands appeared larger, inverted and suspended in the air. A more common experience of this phenomenon is the mirrored reflection of the blue sky on a dark paved road on a hot summer day, giving the appearance of a body of water.

This bit of trivia may help someone win a million bucks on "Who Wants To Be A Millionaire," an amount of money that Sam would have gladly paid for one drink of cool, clear water.

Dan Lutz

Let us remain vigilant and proactive in keeping not only our air and water clean but the global air and water quality as well, so we too can save our breath and our children's and grandchildren's breath.

🍃 THE AIR WE BREATHE

Many areas of the country boast about having a very healthy environment with enough clean water to drink and fresh air to breathe. Still, we must stay vigilant to the quality of our water and air, as well as the environment in other parts of the world. Concerning air quality, areas of clean air can be vulnerable to the airborne legacy from large upwind metropolitan cities and even other countries, regardless of distance.

Dust from the Sahara Desert in Africa can be seen on satellite pictures tracking across the Atlantic Ocean, contaminating the air across the Southeastern region of the United States. Satellite imagery suggests that pollution from Asian industrialized cities carries across the Pacific Ocean, directly impacting the air quality along the West Coast.

In the words of Tuco from the movie, "The Good, the Bad and The Ugly," while setting out across the desert, "If you save your breath, I feel a man like you can manage it. And if you don't manage it, you'll die. Only slowly, very slowly old friend." Let us remain vigilant and proactive in keeping not only our air and water clean but the global air and water quality as well, so we too can save our breath and our children's and grandchildren's breath.

Some Ozone are Good, Some are Downright Ugly

Ozone gas fascinates me because of its multiple personalities: the good, the bad, and the ugly. In this case, the good ozone protects our health, the bad ozone can quickly deteriorate our well-being, and the ugly nature of ozone can destroy the beauty of our natural environment.

The word "ozone" prompts much confusion because of this good, bad, and ugly trichotomy and the fact that, chemically, all three are the identical element. Location, location, location distinguishes the good from the bad and the bad from the ugly.

The good ozone resides on the high end of the ozone-location meter, in the upper atmosphere, about 15 miles above the surface of the Earth. There it continuously filters the sun's harmful ultraviolet radiation, reducing the amount of that cancer-causing radiation from reaching the Earth. On the lower end of the ozone-location ledger, high concentrations of this very reactive form of oxygen can also form near the ground to harm people, animals, plants, and inanimate materials.

This ground level bad ozone can cause shortness of breath, coughing, wheezing, headaches, nausea, eye, and throat irritation, and lung damage. On days when ozone levels are high, it puts children at increased risk for respiratory problems because children breathe more rapidly and inhale more pollution per body weight than adults. People suffering from lung disease can have more trouble breathing when the air is polluted. The ugly consequence of this surface-born ozone slowly eats away century-old buildings, statues, and other historical monuments, and can cause an unpleasant brown haze across the landscape.

Ground-level ozone, as the main component of smog, stands as the most pervasive ecological problem in the industrialized world. Some of our nation's largest cities, including New York, Chicago, Los Angeles, Houston and even the Mile-High city of Denver, often experience high ozone-contaminated smog concentrations, triggering air pollution alerts. These alerts warn people, old and young alike, to cut back outdoor activities.

Big city ground level ozone-contaminated smog (Google Images)

Far away sources do not typically introduce ground-level ozone directly into the big city's air. Instead, sunlight-induced chemical reactions among human-made and organic emissions create near-ground ozone. Carbon and nitrogen dioxide emissions from gasoline engines and industry, as well as biological emissions from trees and other vegetation, interact with sunlight to create the bad ozone close to the ground where we live and breathe.

Ozone is a colorless gas comprised of three oxygen atoms. It's formed in the lower levels of our atmosphere from the reactions of organic gases (RO_2) and oxides of nitrogen (NO_2) in the presence of sunlight and stagnant air producing a radical oxygen atom (O) that reacts with oxygen (O_2) forming ozone (O_3). Organic gases, also known as hydrocarbon-based compounds, are mostly released into the atmosphere by automobile and industrial combustion and solvent evaporation. Combustion from industry also releases oxides of nitrogen into the atmosphere.

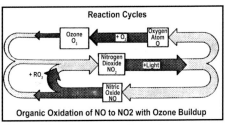

Ground-level ozone reaction cycle

Meteorological conditions, including wind speed and direction, solar radiation, humidity, and temperature, play a significant role in how fast and how large the concentration of ground-level ozone can develop. Specific summer daytime weather conditions can and do produce excessive amounts of ozone near the ground most quickly. In brief, big city morning traffic produces copious amounts of carbon and nitrogen dioxide emissions into the air. Now mix in summertime high temperatures, intense sunshine, weak winds, and a near-ground capping temperature lapse rate with warm air over cool air. From this all too often morning scenario, ozone can quickly accumulate to unhealthy levels in many of the nation's largest cities.

Air Quality Concerns Go Beyond Just Ozone

Many rural regions relish clean air without ground-level ozone concerns; however, even those countryside areas should not take the fresh air they appreciate for granted. Given the right set of circumstances, those sunny days with bright blue skies can almost overnight become lousy air days with hazy brown or smoky skies.

Back to back calm, clear and cold winter days, as described by the weathercaster as beautiful, sunny, crisp days, can lead to steep morning temperature inversions (when the temperature is steady or increasing with height) trapping dust, smoke and soot in the shallow space near the ground and ultimately ingested into our lungs.

Air pollution comes in all sizes and disguises, ranging from invisible gases to highly visible substances. Ozone, one of those invisible gaseous pollutants, can show up as a brown haze over many of our large cities when substantial amounts of the gas, along with high humidity and sunlight, invade the skies overhead. The more visible, albeit tiny, particles of solid or semi-solid material, known as particulate matter, can and do make their way into the atmosphere hovering over many of the country's rural areas. The larger of the small particles include dust and soot, while the smallest of the small particles include smoke.

Dust from man's construction activities (Google Images)

Natural sources of this particulate matter include volcanic ash, pollen, fire and dust blown about by the wind. Human-made sources include dust from roads, freshly plowed fields, crushing and grinding operations, and construction activities. The use of wood in fireplaces and coal in stoves produces rather significant amounts of the smaller particles in localized areas.

The effects of these airborne particles range from visibility degradation to vegetation damage to climate change. The smallest of particles (about seven times smaller than the width of a human hair) can deposit themselves deep in the lungs, only to be trapped by its membranes. When breathed into the lungs, they can cause excessive growth of fibrous lung tissue, leading to permanent lung damage. Children, the elderly, and people suffering from heart or lung disease, find themselves particularly at risk.

Many industries across the land have made significant changes in the way they operate and control the release of human-made dust. Changes in street sanding, parking lot paving, and oversight of construction activities, also make a difference.

Citizens also can do their part to help maintain good air quality. The simple act of burning clean, dry wood can make a noticeable difference in localized air quality. Smoldering fires from burning damp timber, can produce high levels of smoke, especially during periods when temperature inversions trap the smoke near the ground.

Smoke levels typically worsen in the winter due to an increasing number of wood-burning and coal-burning fires. The wintertime atmosphere's ability to form a lid on vertical mixing caused by a deep cold layer of air near the surface with a warm layer of air sitting on top can lead to concentrations of unhealthy levels of smoke-filled corrupt air. Therefore, remain mindful of mornings with steep temperature inversions, particularly during the winter months, when firing up the fireplace.

Daytime heating eradicates this near-ground warm air over cold air trapping inversion, allowing vertical mixing of the air to begin. This up and down mixing then carries the surface pollution, such as dust, smoke, and soot, up and away to its eventual demise at the hands of nature's highly efficient air scrubber known as rain or snow.

Dust Simply a Fact of Life

We breathe it, eat it, sleep with it, cough it up, and spend way too much time trying to wipe it away. It's even on my computer screen as I write this book. There is no end to dust. It's a fact of life on this planet. Sooner or much later, almost everything on Earth becomes dust. The second law of thermodynamics about order versus chaos dictates that all nicely ordered vegetable, animal and mineral objects must eventually break down and become a source of dust.

Nature alone kicks up vast quantities of dust. Volcanoes and oceans convert rocks and saltwater to an over-abundance of atmospheric dust. And of course, let's not forget about the enormous amounts of pollen, spores, and bacteria that blow into the air, adversely affecting our sinus cavities.

Man's activities can also stir up large amounts of dust. Manufactured metal products, such as bridges, buildings, automobiles, etc., begin breaking down into tiny metallic dust particles due to the presence of atmospheric water causing chemical reactions to take place.

In 1862, the Homestead Act opened the Midwest plains and grassland prairies to the farmer's plow. During wet years, abundant crops and deep-rooted grass protected the topsoil. Not so during the consecutive dry years of the Dirty '30s, when wind and drought formed massive clouds of dust, carrying tons and tons of the Great Plains' topsoil thousands of miles away. Today, many farmers and ranchers utilize proven soil conservation measures to protect and prevent topsoil from blowing onto the neighbor's property and into the neighbor's home or lungs.

Satellite image of the Sahara Desert's dust traveling across the Atlantic Ocean (Yahoo Images)

Industrial activities in large cities and extremely arid landscapes can produce large amounts of unhealthy dust. And these extremely fine particulates pay little attention to the county, state, or international borders. The Northeast region of the United States has a difficult time keeping its air dust-free because prevailing southwest winds continuously import dust from not only nearby large industrial cities but areas as far away as Oklahoma and Texas.

Satellites confirm clouds of international dust invade this country's Southeastern shores from lands as far away as Africa's Sahara Desert. Due in no small measure to the Northern Plains' prevailing northwest winds flowing over the mostly fertile rural country, air pollution from upwind dust sources has a minimal impact on air quality here in the Black Hills.

Many cities have drafted plans that outline policy and procedures to reduce dust emissions from city property, city operations, and city-controlled privately-owned operations. When forecasts call for weather conditions to push atmospheric dust to possibly unhealthy levels, local governments issue dust pollution alerts to the public and ask industries to reduce dust-causing operations voluntarily. Also, dust measuring instruments can be found throughout many large cities to monitor daily dust levels, ensuring the air we breathe remains well within safe levels.

On a more positive note, atmospheric dust brings with it two essential and worthwhile physical properties: its ability to scatter light of short wavelengths and its ability to serve as nuclei for the condensation of water vapor. (Verma, 2015) Mist, fog, clouds, rainstorms, and snowstorms would hardly ever occur without the presence of dust particles in the air. Finally, without some form of dust pollution to break-up further the sun's white light, the sunrise and sunset displays, would have little color variation and thus, less beauty and spectacle we've come to expect from the tapestry of contrasting colors, typically blues, violets, and reds.

Even Blue Sky has Science Behind It

After a prolonged steady rain, when the storm clouds finally part, the sky above typically appears bluer than blue. Could it be that we've become accustomed to the sky's day in and day out pre-storm pale blue, making the post-storm blue sky appear especially vivid by comparison? After all, if long-term absence makes the heart grow fonder, why shouldn't the clearing skies take on a more profound and colorful intensity of blue, if only by the perception of those atmospherically challenged observers?

Enough flowery prose! I am a scientist, and, damn it, science has no room for accepting the beauty of the moment without hypothesizing, dissecting, testing, proving and reporting the "what and why" of nature. I must dispel meteorological myths, dispensing with the romantic foolishness that treats physical theories as a function of the heart. Here is the truth: A bluer sky typically follows the rainstorm, and it happens with total disregard of our anticipation of witnessing a return of nature's colorful tapestry after an extended absence.

Selective scattering of the sun's white light by air particles off their intended course spreads the deep blue tapestry across the sky. The particles that scatter light can be air molecules, dust particles, soot, water droplets, gases, or most other particulates.

Also, selective scattering can have the opposite effect by taking away the vibrant blue color of the sky, leaving an unappealing hazy sky. Light scattering by smog forms a particularly unattractive brown haze over some large cities, such as Houston or Los Angeles.

Selective scattering (or Rayleigh scattering) occurs when individual particles are more effective at scattering a specific wavelength of light. The plentiful molecules of air that surround us, like oxygen and nitrogen, selectively scatter out the shorter wavelengths, or the blue light, from the sun's bright sunshine. This selective scattering of the shorter wavelengths gives the sky its blue color we often see on bright sunny days.

Typically, days before a storm, particles of all sizes have had time to fill the sky, scattering light of many different wavelengths hither and yon, culminating in a blur of color somewhere on the blue end of the color spectrum. Before the storm's arrival, the relatively dirty atmosphere can indeed offer up a blue sky, but not the intense blue.

An intense blue sky after snow storm

Now enters the storm bringing either numerous raindrops or snowflakes. The precipitation from these storms does an exceptional job of scrubbing the larger particles out of the air, leaving behind mostly the air molecules to paint the sky the deep crisp blue so commonly seen after a good rain or snow.

The blue-sky theory has a South Dakota connection. The British physicist John Tyndall was the scientific grinch whose approach, the Tyndall Effect, explained why the sky is blue. Nineteenth-century settlers in Bon Homme County, South Dakota, named their county seat Tyndall, to honor the man who helped to remove the romance from the color of the sky.

Morning Air Temperature Traps Plant Smoke

Catastrophic fires place at risk not only firefighters but the public as well. The harmful effects created by these sizable fires, whether massive forest fires or fires that destroy large structures, can often spread quickly beyond the immediate bounds of the inferno. In no small measure, the dispatch and degree of the fire's impact on the surrounding community depend on winds and outside air temperature.

On January 30, 2002, a massive structure fire erupted that destroyed a large meatpacking building located in Rapid City. Light westerly winds carried the smoke from the blaze eastward toward the heart of the city. Knowing the likely movement of the smoke, coupled with the fear that a 1,500-gallon tank of anhydrous ammonia might also release toxic fumes downwind, the Pennington county Emergency Management Office issued an alert warning to the downwind public to remain indoors.

Of immediate concern, the officials managing the emergency operations needed to know wind direction and speed to help make informed decisions; also, the ambient air temperature played a prominent role in issuing the alert warning.

The fire became fully involved with heavy smoke during the early morning hours when the day's coldest surface ambient air temperature occurred. That morning's cold surface temperature, layered beneath warmer air aloft, set up a steep surface inversion that prevented the smoke from rising skyward. Instead, the inversion trapped the toxic smoke in a stable shallow layer near the ground.

Here's how the temperature sequence played out that morning: The vertical temperature of the air near the surface cools typically with height. The higher one goes within the troposphere, the colder the temperature. This temperature configuration of relatively warm air under cooler air aloft allows for up and down currents of air to mix high concentra-

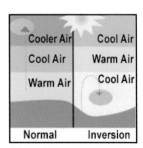

Near-surface temperature inversion (Yahoo Images)

tions of atmospheric pollutants and, in this case, smoke. Under these quasi-turbulent conditions, the toxic smoke from the fire would have been carried aloft, within a very short distance, and dispersed far-off from our lungs.

Not the case during the morning hours of this structure fire when nighttime air temperatures near the ground cooled off drastically under a cloudless sky. Because the air temperature cools the greatest near the ground at night, when compared to the air temperature aloft, for at least a few hours during the morning of the fire, the near-surface air warmed with height. As a result, a temperature inversion or a highly stable layer of air developed near the surface.

This reversal of the standard vertical temperature profile places a lid on any vertical air currents, thus trapping pollutants within a relatively shallow layer of air near the surface. In Rapid City, on that morning, the smoke released by the structure fire remained trapped within a relatively thin layer of air at the surface, not only near the fire but for some distance downwind. Classes at the nearby Rapid City Central High School were called off as a precaution.

Only during the afternoon, when the sun had time to warm the surface and the near-surface air, reversing the effects of the temperature inversion, did the smoke rise and dispersed at higher levels. An understanding of both the speed and direction of the horizontal winds and vertical temperature profile determined when and for how long to issue the cautionary air-quality warning to those people and businesses located downwind.

Fires and the Bermuda High Take the Blue Away

The summer's heat in the Black Hills region peaks during the first few weeks of August. Along with those high afternoon temperatures, the dog days of the 2017 summer appeared more hazy than usual, with a profusion of contaminates in the air.

This overabundance of air pollutants settled over the Black Hills area for a good reason. You see, 2017 happened to have been one of the worse years for forest fires across the Western United States. Numerous wildfires raged across California, Utah, Montana, Nevada and Idaho, generating copious amounts of billowing smoke to fill the heavens and blur the sky's natural blue color for hundreds of miles downwind.

Given the typical configuration of the summer's upper-level steering winds, the smoke from these immense fires traveled northeastward from California, Utah, and Idaho toward Montana where the now Montana winds carried the accompanying smoke southeastward toward the Dakotas.

These overhead steering winds effectively carried the smoke and its filtering effects onto the Northern Plains. The ordinarily deep blue skies over the Black

Hills then appeared hazy, which tended to add imaginary degrees of hotness to one's perception of the day's already sizzling high temperature.

One other significant ingredient, besides the wind and forest fires, can be found in our smoky mix. It's necessary to stir in a chunk of the Bermuda High-pressure system, which spends its summer residing over the Western Atlantic, to complete our smoggy recipe. Because of this semi-permanent high-pressure weather system, the Eastern United States, including Eastern South Dakota, endures extended periods of hot and muggy summer days.

In some years, an elongated piece of the Bermuda High can reach well into the Western States, including the Black Hills, causing unusually dry and hot weather conditions. More importantly, this vast mass of stagnant air prevents any natural vertical venting by the atmosphere which traps the smoke near the surface and into our eyes and lungs.

Elongated Bermuda High pressure system expanding into the Northern Plains

As discussed earlier, the vertical temperature in the tropospheric layer of the atmosphere decreases typically with height. The higher one goes, the cooler the temperature. This unstable cold air over hot air temperature configuration allows the air to vertically exhaust high concentrations of low-level atmospheric pollutants, such as smoke and dust.

That's not the case when an area of a robust high pressure, such as from the Bermuda High, settles overhead. High-pressure systems compress and heat the air using the same principles as a pressure cooker. This compression process can cause a shallow layer of air near the surface to warm with height, creating a reversal of air temperature from the standard cooling temperature lapse rate with height.

This reversal of vertical temperature, now warm over cool, places a lid on any vertical air currents necessary to exhaust the forest fire smoke from near the surface. The smoke released by the too often and too numerous summer wildfires across Western Canada and the Northwestern regions of the United States finds its way to the Black Hills only to remain trapped within a relatively thin layer of air near the surface for lengthy periods.

Once the Bermuda High backs off and loosens its grip on the Upper Plains, the deep blue sky returns, along with seasonal temperatures, lower humidity and our breathtaking view of the Black Hills.

Dan Lutz

Where differences in the waves of air pressure exist, there is a constant striving by the forces of nature to balance them. I believe that such diversity found in these waves of pressure and their impact on the goings-on of nature maintains and, at times, intensifies our deep and abiding interest in weather.

🍃 THE AIR PRESSURE ALL AROUND

I believe the inspiration for many creative endeavors emerges from the simplest of life's experiences. In my case, nature's surroundings dictate the subject matter in many of the writings found in this book.

The mere formality of walking my dog, "Wally," through the park will often allow my mind to wander into the realm of weather. I'm free to contemplate the park's surroundings while Wally tracks down that strange smell wafting about his nose on the day's prevailing wind, or hunts down that rustling sound among the bushes, or chases after the sight of that squirrel scampering up a tree. These many strolls through the park often inspire my thoughts about the essential and vital role weather plays in nature.

The particulars found in the park's surroundings, such as cloud shapes and their movement overhead, the splashing of Rapid Creek over and around smoothly worn rocks, and even the various sounds of a woodpecker striking a dead tree, can also explain the weather. You see, like Wally, we sense our surroundings through waves traveling through the atmosphere such as sound and light waves. We also sense weather's presence through waves of a different persuasion, waves of pressure passing through and interacting with the atmosphere. Pressure waves cover a spectrum of scales from triggering a slight puff of warm air to generating periodic bands of clouds drifting across the sky later intensifying into mamoth thunderstorms, to dimensions worldwide, causing the peaks and valleys of the jet stream stretching around the globe.

These pressure waves of weather, like the ocean waves splashing on the shore, differ notably in size and frequency and fluctuate significantly from day to day and even from moment to moment. No two weather events are ever exactly alike because pressure waves can vary into an infinite number of combinations as they build up and travel throughout the atmosphere only to waste away later.

Clouds grow and drift across the sky and disappear, never to repeat the same appearance. The park's setting that reeled from blizzard conditions today may enjoy sunny and mild conditions tomorrow. On an even larger scale, last year the park may have experienced a dry spell, while this year it may receive more than ample moisture — all due to waves of pressure playing games with our atmosphere.

Where differences in the waves of air pressure exist, there is a constant striving by the forces of nature to balance them. I believe that such diversity found in these waves of pressure and their impact on the goings-on of nature maintains and, at times, intensifies our deep and abiding interest in weather. I know that every walk through the same park along the same trail with the same dog will not be the same as with previous strolls. Those sights, smells, and sounds will not be the same. The hourly, daily, weekly or yearly dynamic variations of the weather we sense, too, will not be the same, keeping weather the leading topic of discussion at home, in coffee shops, on social media or in the office around the water cooler.

The Highs and Lows of Pressure

Let me now explain those blue Hs and red Ls that slide across the weather map, marking transitions from current weather to tomorrow's forecast. In meteorology, the H on the weather map does not stand for high temperature, but instead, high pressure. Similarly, L does not indicate low temperature but low pressure.

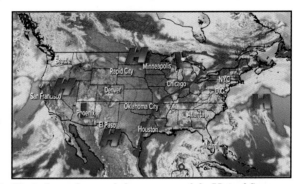

Typical weather symbols on a map of the United States

Approximately 400 miles of air piled up over our heads produces a significant pressure force on our bodies, and on the Earth as well. Typically, we do not feel air's pressure (about 18 pounds or the approximate weight of a bowling ball) on our bodies. You see, the air pressure inside our bodies pushing out equalizes the air pressure outside our bodies pushing in. The push-in and push-out pressure forces keeps our bodies in comfortable balance with the environment.

You will feel, however, abrupt pressure changes on your body like when your ears pop when taking off or landing in an airplane or speeding up and down those steep mountain roads. The plane makes a rapid transition from high pressure to lower pressure as it takes off and vice versa as it lands. It's the inside body pressure trying to get outside as the plane climbs to heights of less pressure and the outside air pressure trying to get inside our body as the plane lands that causes the ears to pop.

In the Northern Hemisphere, the wind blows clockwise around the blue H (High pressure) on the weather map and counterclockwise around the red L (Low pressure) on the map. Remember those two rules-of-thumb and the whole weather map should make more sense. When outside, turn your back to the wind and lift your left arm out to the side and it will point in the general direction of the center of the low-pressure system. Of course, if you move to the Southern Hemisphere, the formula reverses; but that's a subject for another time.

The blue Hs also represents the center of large globs of air which have a reasonably uniform distribution of temperature and moisture. We call these globs of high air pressure "air masses." Though weather charts depict air masses as two dimensional, they are three-dimensional, i.e., spread out north to south, east to west and bottom to top in the vertical, and take up a wide-ranging portion of the lower atmosphere.

To become an actual air mass of high pressure, it must stay in one location long enough to develop the characteristics of the region. Two blue Hs on a weather map separated by a front, either cold, warm or stationary, can represent two distinctly different air masses often from different geographic areas.

Air masses even have names to designate from where they originated. Here on the Northern Plains, we concern ourselves with four air mass varieties: continental polar (originating over the dry Arctic region), maritime polar (originating over the moist Northeastern Pacific Ocean), continental tropical (out of the arid Mexican Plateau), and the maritime tropical (starting out over the moisture-rich Gulf of Mexico, Caribbean or the South Pacific).

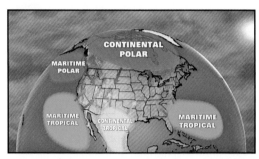

Typical locations of the North American semi-permanent air masses

Continental polar air masses, originating over land, usually have low water content. These air masses slide across the Northern Plains, and they typically provide the region with low humidity and clear skies. The maritime air masses, originating over vast bodies of water, regularly contain a higher water content. Most often, the common Northern Plains maritime polar air masses bring clouds, high humidity, and rain or snow to the Black Hills region.

The cold fronts that invade the Northern Plains in the winter usher in either continental polar or maritime polar air masses. During the summer months, the warm fronts sliding into the Northern Plains frequently usher in the maritime tropical air masses. The less humid continental tropical air masses reach the Northern Plains only occasionally.

After a long winter of below freezing temperatures, South Dakotans eagerly welcome the arrival of a warm tropical air mass. The wait for the arrival of the warmth may seem like an eternity, but in fact, those tropical breezes can occur quickly under the influence of an expanding sub-tropical jet stream reaching northward out of the Southeast Pacific Ocean, Mexico, or the Gulf of Mexico and into the Northern Plains.

So, if you seek warmth, you can talk to a travel agent about a trip to Cancun, Mexico. Or, for a more cost-effective approach, you can wait for the sub-tropical jet stream and let Mexico come to you. Ole!

Thick Air Masses Key to a White Winter

Often, snow-loving residents of the Black Hills region eagerly anticipate significant snow but from time to time end up with only a dusting of the white stuff, or a disappointing cold rain. On many of those mornings when the skies were gray with thick water-laden low clouds blowing in from the northwest and temperatures near the ground hovered near the freezing mark, the locals may anticipate significant snowfall to follow and yet snow did not materialize.

A snow-covered winter day in the Black Hills, SD

On those days when the wintry mix suggests significant snowfall, but the snow comes up woefully scarce, the blame lies not so much with the lack of moisture in the air, nor air temperature. On those days with low snowfall amounts, the charge may rest entirely with "thickness." You see, those high-pressure domes of cold air that pass over the Black Hills have depth or thickness. And it's the thickness of those chilly high-pressure areas that sets the stage for either a notable snow event or a mere dusting.

Cold air plunging through the Northern Plains demonstrates some of the same characteristics as a truckload of grain poured out in an elevator. The grain, when it hits the ground, spreads out in all directions, and the farther the grain spreads, the shallower the depth at the edge of the pile.

Typically, mounds of cold air, originating over Siberia, pour out of Canada and spread their "thick" layer of cold air across the Northern Plains, while moisture-laden air from the Gulf of Mexico traveling northward rides over the top. The lifting of the Gulf air by the denser cold Canadian air allows the Gulf air to lose its stored-up moisture in the form of precipitation.

The depth or thickness of the cold dome over an area determines whether the precipitation is rain, freezing rain, sleet or just snow. Often the depth of the cold dome changes considerably within the expanse of a few miles, which can produce fine snow in Sturgis, SD, sleet in Rapid City, SD, freezing rain over Ellsworth AFB, and rain in Wall, SD.

How the various types of precipitation form from different thickness of a cold dome of high pressure (freezing air)

Sleet can be as common as snow across the Northern Plains. Sleet, also known as "ice pellets," occurs when rain from a relatively warm layer of air falls through a deep layer of subfreezing cold air near the ground. As the raindrops pass through the lower and colder layer of air, they quickly freeze into grains or pellets of ice hitting the ground before any further transformation can take place. Sleet is a type of precipitation consisting of transparent or translucent fragments of ice, 0.2 inch or less in diameter, that bounce when hitting the hard ground and may make a sound.

Closely akin to sleet is freezing rain, which is rain that falls through a thinner layer of below-freezing air located close to the ground. The supercooled rain drops then freeze upon impact, forming a coat of ice or glaze on the ground and other exposed objects. Freezing rain is wedged between the occurrence of rain and sleet.

The impacted surface must be at or below freezing to have freezing rain, and furthermore, the raindrops themselves must be supercooled. Supercooled means that the temperature of the liquid drop is below freezing. Technically, 32 degrees Fahrenheit – or if you prefer, zero Celsius – is the melting temperature of ice and not the freezing temperature of the water. A pure water drop will remain liquid at temperatures well below 32 degrees Fahrenheit. When temperatures reach 40 degrees Fahrenheit below zero, all liquid water will freeze.

Along with freezing rain comes the hazardous driving condition known as "black ice." Black ice can trick a driver into thinking it's water on the road ahead when, in fact, it's an extra thin layer of ice. Black ice is likely to form in shady spots on the road, under bridges and overpasses.

To get a significant amount of snowfall on the ground, the source of the flakes must originate within a substantial depth of cold air. On those days when you find minimal snow accumulation or a cold rain at your feet, you can bet with some confidence, that you're standing near or under the shallower periphery of the cold dome of high pressure.

The winter season orientation of the La Nina polar jet stream

When during the winter season the polar jet stream fails to push those thick cold domes of air out of Siberia and far enough into the Northern Plains, snowfall may be lacking. And if you wish to stray one step farther along the chain of weather events to seek out the cause of many snowy winters, pay attention to the La Niña years. La Niña strongly influences the polar jet stream, to push thicker Arctic air farther southward, bringing colder and snowier winters to the Northern Plains.

Pressure on at Turkey Time

Let me offer a quick weather truism that might come in handy when planning the Thanksgiving or Christmas turkey dinner. Heating trapped air causes its molecules to become more agitated, which increases the air pressure within a confined space. This concept not only pertains to meteorology, but to the culinary arts as well. I am not talking about a bigger and better pressure cooker, nor the heat from the baker's kitchen when making caramel rolls. I'm talking turkey.

For some reason, our insatiable fervor to improve everything from shampoo to orange juice to toothbrushes now includes cooking the turkey. My advice to all the turkey technicians out there, beta test your "technologically superior" cooking methods before releasing them to the public, because a terrible flaw exists in the now "supposedly improved" turkey cooking method.

Chef Turkey (Google images)

Let me illustrate by using one memorable Thanksgiving dinner at the Riggio house. The cornerstone of every Thanksgiving tradition, like football, mashed potatoes, gravy, and pumpkin pie is the turkey "timeline." Knowing when the turkey is thoroughly cooked locks in the time when friends and relatives arrive, when to start the mashed potatoes, gravy and string beans. And, it tells you which quarter of the football game you might miss. Folks, planning the Thanksgiving dinner cannot begin without knowing the turkey timeline.

In simpler times, you established the turkey timeline by its weight or one hour for every three-pound turkey weight. A 12-pound turkey cooks for four hours. Put it in at 11 AM, it's cooked thoroughly at 3 PM, and dinner is served at 3:30 PM. Make all other plans accordingly. Simple and straightforward.

But, that's much too elementary for the turkey technicians. Now, impaled in the turkeys one can find this little doohickey that pops out suggesting the bird is fully cooked. In effect, the technicians gave control of the Thanksgiving meal, and if you are not careful the Christmas meal, to the turkey. No longer do you know the turkey timeline when placing the bird in the oven. You now find yourself stressing out while waiting for the turkey to push that little red stick out.

Unfortunately, the Riggio Thanksgiving turkey had one of those pop-out sticks. Using the old cooking math based on weight, I calculated removing the turkey from the oven around 2 PM. As the clock on the wall displayed the big hand on the 12 and the little hand on the 2, the turkey looked done, but the pop-out thingamabob had yet to pop out, creating quite a bit of anxiety and doubt on my part. The apparent dilemma: Do I trust the old way of timing the turkey or this new technologically advanced pop-out device?

With visions of botulism and doubled-over dinner guests dancing through my head, I waited and waited for the red pop-out gizmo to do its thing. Guests were hungry, gravy not made, mashed potatoes getting cold, and the game going into overtime, all because those anonymous turkey technicians gave control to the dumbest of all birds, the turkey.

After waiting for one full half-hour beyond the "by weight" cook time, I decided to pull the bird out of the oven and carve the bird even with the pop-out doodad firmly in its original state. To my relief, once the bird left the oven, the thingamajig popped out.

With Christmas dinner not too far away, you might wish to attach this section to your refrigerator door as a reminder. I can only conclude that the heat caused the pressure in the confined oven space to increase, preventing the thingamabob from popping out at its allotted time. Once the oven door opened, the oven pressure lessened by equalizing with the outside pressure, allowing the red gizmo to appear.

To lessen other's frustrations, I propose that all turkeys with the pop-out thingamajig include a label warning the cooks about this simple weather concept: "When air is trapped and heated, its pressure increases."

Downhill All the Way

Years ago, when my family and I would hit the ski slopes, my novice kids skied the bunny slopes; while I, the more seasoned skier, hit the bunny slopes, as well. I rationalized the less risky choice of trails with the phrase, "I like to cruise down the mountain, so that I can enjoy the passing scenery." I would tackle the more challenging blue-marked slopes with ever-growing confidence later in the season, but never did I voluntarily ski the near free-fall of the black-diamond hills. I like slow, because "slow is smooth and smooth is fast."

I've been a fan of slow all my life. Slow allows time to think, to talk, to get to know. The black diamond-marking on some slopes implies pure speed; because, on those slopes, gravity has nearly a free hand over your body and at times your will. If it were possible to enhance the friction force to help maintain slow as gravity force increases, then I, the proponent of slow, could then claim the steep black diamond slopes as my own. And, how I would secretly love to brag about my hair-raising downhill runs with names like "Steep Doom."

To introduce slow to the black diamond slopes, we need to understand how skis work on snow. Skiing involves contact between two surfaces, the bottom of the skis and the snow, plus the concept of friction. Friction is the property that objects have, that makes them resist movement across one another. There must be minimal friction between the skis and the snow to glide quickly down a mountainside. Therefore, it stands to reason that increasing friction between my skies and the snow should allow me to ski down the steepest slope at my own pace.

Some researchers claim that snow becomes more slippery with less friction when the surface partially melts under pressure (increased pressure increases heat), causing a thin layer of water to develop at the point of contact. If the snow is too cold, that pressure will not be able to melt the thin layer of snow. Without the slight melt, more friction occurs, causing what I call "slow snow," slowing the skier. This theory suggests that for the fastest skiing, the temperature must be cold enough for snow to form, yet warm enough so that the skier's weight can apply enough pressure to heat the snow to create the thin layer of water.

Downhill skiing, Terry Peak, SD

Will "slow snow" ever allow me to hit the black diamond slopes? Probably not. Gravity will always rule the slopes. Most skiers enjoy the thrill that goes along with the speed, the jumps, and the moguls of the black-diamond grades. I will content myself watching the pine trees glide by as I cruise the bunny hills and eventually work up to the blue-marked slopes.

At least then, while cruising the non-black diamond slopes, I will have time to dream up some tall tale about my encounter with "Steep Doom" that I can tell over a cup of hot chocolate.

East Meets West in Foggy Gap

Located here in Rapid City are two hills separated by the Gap. The hill to the north is Cowboy Hill, and the hill to the south is Skyline Drive. The Gap lies in between through which morning and evening traffic travels.

On occasion, as a regular morning Rapid City commuter, you have undoubtedly encountered a thick curtain of fog while passing through the Gap. You were either moving onward into the fog or onward out of the fog depending on which direction you were traveling.

When heading east, you would be slipping into a dense gray fog shrouding both the sun and other commuters three car lengths ahead. When heading west, you would be moving out of the foggy mist toward the sunlit blue sky and a completely different weather day. This quasi-foggy weather typically lingers in the Gap and across East Rapid City until four to five hours after sunrise, after which the sun's illumination spreads a more consistent brilliance across the entire city.

Fog in the Gap, Rapid City, SD

Meteorologically speaking, it's "upslope fog" that envelops the Gap and most of East Rapid City on those days. Upslope fog alone may not seem so unusual except for the sharp line, every bit as distinguishable as the dotted line on a Wheaties cereal box top, separating fog from clear. Once you crossed the line that morning from fog to sunshine, your ability to see certainly changed and possibly your mood, too. Of course, traveling in the opposite direction, you can easily experience mood reversal.

Those of us living in and around Rapid City understand that the term "upslope" usually implies inclement weather. I often use "upslope" in my weathercast, because of its significant contribution to our frequently changing weather along the front range of the Black Hills.

Distinct seperation between upslope fog and clear air in Rapid City, SD

With a low-pressure area located to the south of the Black Hills and high-pressure area to the north, the wind pattern over West River blows upslope from the lower plains to the east to the higher foothills and, eventually, to the even higher Black Hills to the west. Upslope easterly winds will bring wet and cooler weather to our region more often than not.

As the air rises to lower pressure at higher levels, it expands and cools. If the air cools enough, the water vapor gas in it condenses to fog, clouds, rain, snow, or some combination. Rising air caused by upsloping terrain cools at the rate of about 5.5 degrees for every 1,000 feet it rises. When the air cools to the dew point temperature it condenses, changing from gas to liquid and often bringing widespread fog to the Eastern Black Hills foothills and rain or snow to the higher elevations. The same happens in most mountain ranges, often accounting for more snow or rain in the mountains than in the nearby lower elevations.

This same upslope principle holds in the Gap, albeit at a smaller scale. When a moisture-laden easterly wind blows across Rapid City, upslope fog can occur in and east of the Gap. West of the Gap that same wind moves slightly downslope and sinks reversing the thermodynamic processes causing the air to warm and dry as it moves down from Cowboy Hill and Skyline Drive into West Rapid City abruptly clearing away all remnants of fog and possibly an awful disposition.

This up or down air flow causes notable weather differences between west and east Rapid City, often clearly visible in the Gap on not so clearly visible mornings.

Meg Roman

Only when temperature fields exist in an ordered fashion, i.e., hot in one region, and cold in a neighboring region, can the second law of thermodynamics do its thing and transfer energy from the hot to the cold, which in turn makes the weather happen across local, regional, national and even seasonal scales.

🍃 THE TEMPERATURE WE FEEL

South Dakota has a pleasant middle of the country type of climate. On average the summers can be somewhat humid and warm, while the winters dry and cold.

Goldilocks remarked that, "This porridge is too cold, this porridge is too hot, but this porridge is just right." And by just right, she probably meant warm. The hot soup mixed with the cold soup resulted in a "just right" warm soup. The heat or energy stored in the hot soup transferred to the cold soup warming it up, allowing Goldilocks to eat her just right soup.

In the weather business, we experience too hot temperatures, too cold temperatures, and just right temperatures, all critical ingredients when it comes to weather. Above average temperatures and below average temperatures demonstrate order exists within the weather's temperature fields. Ordered temperature fields are necessary if we like rain, snow, wind, clouds and the rest of the weather paraphernalia.

Without ordered temperature fields, or in other words, with only "just right" temperatures across the globe, weather cannot happen. Only when temperature fields exist in an ordered fashion, i.e., hot in one region, and cold in a neighboring region, can the second law of thermodynamics do its thing and transfer energy from the hot to the cold, which in turn makes the weather happen across local, regional, national and even seasonal scales.

As discussed earlier, because the Earth is round, tilted and the sun's energy comes from one point, some regions receive much more heat than required, while other areas receive less than needed heat allowing for the transfer of energy/heat and the making of weather as we know it. Earth wants to be in balance, but the sun's energy contribution puts it out of balance – hence the daily battle between the always present too hot and too cold imbalance results in the regular transfer of energy, giving us all sorts of weather.

Average Annual High/Low Temperatures
(National Climatic Data Center)

East River Communities	Annual Low Temperature (°F)	Annual High Temperature (°F)	West River Communities	Annual Low Temperature (°F)	Annual High Temperature (°F)
Aberdeen	31.5	54.5	Custer	32.4	56.1
Brookings	32.5	53.8	Deadwood	31.4	55.8
Huron	34.4	57.4	Hot Springs	34.0	63.6
Sioux Falls	34.8	56.4	Lead	33.9	54.8
Vermillion	37.8	60.6	Philip	32.8	60.8
Watertown	32.5	53.6	Interior	38.1	63.8
Yankton	36.0	60.2	Rapid City	33.5	59.1
Pierre	35.4	59.0	Hill City	31.3	58.7
Mobridge	34.5	56.3	Lemmon	33.3	57.4

Extremes Work to Make Weather What It Is

The weather-related question I heard a lot during much of the past half-century was, "What's with all the above normal hot temperatures?" or "What's with all the below normal cold temperatures?" A few of the past autumn seasons were some of the warmest ever on record dating back more than 100 years. But even though we found those mild autumn days pleasant, we need the customary colder autumn and winter days.

With colder temperatures, much good happens. First and foremost, freezing temperatures kill many of those summer-loving bugs, and I believe some bugs should die. If not, they come back in the spring with hordes of offspring. After an autumn and winter of romping and frolicking in the warmer than normal West River soil, their higher than normal numbers of offspring return to eating more plants, smear more windshields and, worse yet, bite more people, dogs, and cats. Secondly, many of us enjoy snow sports activities, and you need cold temperatures to get snow.

So, what's with the warmer than or colder than usual weather? Well, I can talk about the extensive ridge of high pressure over the country's midsection that kept the Black Hills' winter storms at bay and temperatures above average. Or the rare sharp dip in the jet stream pushing frigid Arctic air across the Upper Plains. But I won't. What I can tell you is this: Because of one year's record-breaking hot temperatures and the previous year's record cold temperatures, all is right with the world of weather.

Unseasonably hot or cold periods signify a profound condition in the realm of weather that goes way beyond the apparent extension of hot or cold weather outdoor activities. Let me explain by example. An ice cube in a glass of hot water represents an ordered temperature field. Initially, the ice cube contains all the cold or all the slower moving molecules, while the hot water contains all the hot or the faster moving molecules. The order is unmistakable: the ice is cold, and the water is hot.

With a room temperature above freezing, the hot water will always transfer its heat to the now melting ice cube, cooling the hot water to warm water. Just as the second law of thermodynamics instructs, order (cold ice and hot water) gives way to disorder (warm water) through the transfer of energy. So, when temperature fields are ordered, like cold Arctic air masses next to warm tropical air masses, or like our unseasonably hot or cold periods discussed above, the second law dictates a transfer of energy/heat can take place, which in turn makes the weather work.

In my opinion, even in life, to grow and move forward, existing ordered systems break down. In a debate, for example, one group presents a well-researched and well-organized ordered opinion to a second group, with their own well-organized ordered but contrasting view. The energy transferred between the groups as they argue their own carefully researched opinions often results in a workable resolution between the different points of view. And more times than not, the compromise or the breaking down of the two ordered system is the healthier path for continued development and betterment of the whole. Without the transfer of energy between opposing ideologies to invigorate open discussions, progress often comes to a standstill and dies.

An old Texas politician once said, "All you find in the middle of the road are two yellow stripes and dead armadillos." With too hot temperatures and other temperatures too cold, demonstrating order exists within our weather systems, that old politician thoughts could not be more wrong when it comes to debate or more to the point of this book – the weather. Without those extreme temperatures, the weather would die off like that middle-of-the-road armadillo.

Can't Stand Heat? Hydrate

After a summer season's excessively high temperatures, followed by our desire for the more cooling comforts, this almost always triggers the inevitable debate about dry heat versus wet heat. Let's now put it to rest.

We hear it every summer from the folks living in the Southwestern drier climes say, "It doesn't feel like 110 degrees Fahrenheit because it's a dry heat." That cliché, often cited in travel brochures circulating throughout the Southwest, has some truth to it. Keep in mind though, a dry bulb thermometer will always read the same temperature regardless of a 10 percent humidity, a 40 percent humidity or even a 70 percent humidity. The ambient air temperatures stay the same.

A temperature of 110 degrees Fahrenheit in the desert community of Phoenix, AZ, is just as hot as 110 degrees Fahrenheit in the more humid municipality of Miami, FL. Air temperature measures the degree of energy (speed of the moving air molecules) in the air or degree of hotness or coldness of the air on some definite scale. Humidity measures something entirely different; it measures the water-vapor content of the air.

Now, after stating the air's heat and moisture content are two different things, let me say that humidity plays a large part in how the human body feels or reacts, to air temperature. The operative word is "feel." Humidity does not change the air temperature reading, but it does change the air temperature's effect on our body.

During those periods of sweltering weather, the heart, more than any other organ in the body, becomes increasingly stressed. When the body's heat rises, the heart works harder to cool our many different body parts by pumping blood to capillaries found near the outer layers of the skin, allowing the blood's heat to escape to the surrounding air.

When the summer days are mild, the body can readily transfer the blood's heat to the outside air through the outer layers of skin, thus efficiently maintaining the body's temperature at a comfortable 98.6 degrees Fahrenheit. Occasionally, summer temperatures can be excessively hot and any extended period with sweltering temperatures and high humidity can hamper the body's attempt to give off its surplus heat.

Always hydrate during excessive heat (Yahoo Images)

The long-standing second law of thermodynamics tells us that energy/heat moves from the body having the higher temperature to the body having the lower temperature (i.e., hot water in a glass transfers its heat to the ice cube, melting it). Therefore, when an inordinately sweltering breeze much warmer than our body temperature circulates close to us, it's difficult for the relative cooler human body to give off its extra heat to the much warmer ambient air, which can result in heat exhaustion or stroke.

All is not lost, because good old-fashioned sweat and humidity, or the lack of humidity, can bring balance back to the body's attempt to give off heat to maintain its 98.6-degree Fahrenheit temperature. Here's how it works: When sweat evaporates (the water to gas cooling process), it uses up heat from the skin, thus cooling the skin's surface. This evaporative cooling helps to lower the skin temperature, and subsequently, the blood and finally the body's temperature.

With excessively hot air temperatures and low humidity, let's say 30 percent, perspiration can evaporate efficiently, thus cooling the body effectively. However, when the humidity is high, let's say more than 60 percent, sweat evaporates much less efficiently. The now moisture-rich atmosphere can only extract much less water from the skin's sweat, severely reducing the body's evaporative cooling process.

With hot temperatures and low humidity, the cooling of the skin and blood works much more efficiently, and the body feels comfortably cool. High temperatures coupled with high humidity diminish the evaporative cooling process of the skin and blood greatly, leaving the body feeling uncomfortably warm.

Finally, a hydrated body is a sweaty body, so keep your inflow and outflow fluids in balance by drinking lots of water, especially during those scorching days. In that way, your perspiration thermostat will find it much easier to stay in balance and in good working order.

Snow's Complex Structure Gives it a High Albedo

Many years back I announced over the airways that one winter storm would briefly end an extended warmer and drier-than-normal period. I went on to say the next two years will continue the warm and dry trend. Not only did the snow from that storm fall from the heavens big time, but the temperatures dropped well below normal across the Black Hills and South Dakota. Eventually, the abundance of snow from that storm will melt to the delight of almost everybody whose livelihood depends on plants that grow in the springtime and water that flows above and below ground.

With big-time snow on the ground, despite a few second thoughts, I stood resolute, right or wrong, behind my drier and warmer long-range forecast. During the following few weeks, much snow dropped from the sky, forcing many folks to break out the scrapers, snow shovels, snow-blowers, and plows to remove the mutant pin-sized ice crystals from their windshields, sidewalks, driveways, and roads. And, it forced me to rethink my long-range forecast and turn my attention to those invading hordes of snowflakes that covered the prairies, parks, hills, and trees.

Those countless numbers of tiny snowflakes always begin as an invisible, tasteless, always-present gas known as water vapor. Winter's cold temperatures and upward-moving air transform the water vapor into snow-making clouds packed with ice crystals. These now airborne ice crystals take on different shapes and composition depending on the cloud type from which they descend. Snowflakes, the most common type of snow, form

Various snow crystal shapes (Yahoo Images)

in clouds tall enough to accommodate both cold and warm layers. The ice crystals grow in the cold layers and melt partially in the warm layers enabling them to cling onto other ice crystals, creating snowflakes.

Once the snow's Earth-bound journey settles on the ground, the concept of snow changes from a myriad of floating individual snowflakes to a single expansive blanket of white covering the ground and taking on a very different personality and weather-making role. With snow covering the ground, daytime and nighttime air temperature becomes notably colder, and it's not just because of the snow's cooling effect on the air above, but because of a phenomenon called "albedo."

Albedo originated from the Latin word for white. It refers to the amount/percentage of light an object reflects away. Instead of absorbing light, snow's complex structure prevents the light from being incorporated into its lattice formation, thus giving it an extremely high reflectivity or albedo. The countless number of snowflakes found in a bank of snow then quickly scatter a beam of white sunlight away from the snowbank, giving the snow its white appearance. So, although many natural objects get their blue, red, and yellow colors by scattering back only that color – snowflakes contain characteristics that scatter back all the white light.

Snow's high albedo also efficiently reflects away much of the sun's energy needed not only to melt the snow but also necessary to warm the surrounding air. Therefore, when all these billions upon billions of snowflakes cover the ground, the sun alone has a difficult time warming afternoon temperatures, which also leads to colder nighttime temperatures.

Only after the snow has melted away can the sun better do its thing to warm our days and subsequently, our evenings. And optimistically allow my warmer than normal long-range forecast to come to fruition.

Frost — Another Dawn Delight

In general, people fall into two categories — morning people and evening people. I am of the morning persuasion, but I'm married to one of the nighttime doctrines.

One reason for my forenoon fetish is the karma that surrounds the morning coffee. The taste and aroma of morning coffee embellish the dawn's radiance. My Viva La Java Mocha love affair with a warm cup of rich coffee rings particularly true the time of the year when the early-morning hours include a blanket of white feathery frost on the grass, on the bushes and leaves, on the roofs, and on the backyard deck railing.

After a spring and summer of this seemingly bio-illogical cycle of growing, picking, or mowing every week, the arrival of frost can be a welcome sight. So, let's just say, besides enhancing the morning coffee experience, frost is nature's way of ending our annual ritual of meticulous planting, watering and fertilizing various forms of plant life.

When overnight temperatures fall below the freezing point, frost can form over most outdoor surfaces, a deck railing or a blade of grass, for instance, as a by-product of nighttime radiation cooling. With the setting of the sun, the amount of the sun's heating rays quickly decreases, and the outgoing heat from the Earth's surface quickly outpaces the incoming energy. Evaporation of moisture stops, for the most part, and the Earth's surface begins the cooling leg of the daily heat budget cycle.

Significant day to night variations of temperature occurs most often during the fall season when humidity and cloud cover are frequently low. The difference between high daytime temperatures and low nighttime temperatures can easily range as much as 50-60 degrees Fahrenheit in the High Plains. So, an afternoon high temperature in the 70s can quickly give way to below freezing nighttime temperatures and morning frost.

Frost crystals on leaves

Day to night temperature variations can also vary between urban and rural areas caused by the heat island effect. On many nights, changes in the heat budget between urban and rural regions can differ as much as 10 degrees. Temperatures in urban centers are typically warmer than areas at the same elevation located on the outskirts of the city. The vast amounts of concrete, asphalt, and brick found in urban localities store large quantities of the afternoon heat energy, slowing nighttime cooling when compared to the close-by open prairies and farmlands.

Temperatures tend to bottom out at daybreak, the point at which much of the Earth's stored heat from the previous day's solar radiation departed from all my backyard's exposed surfaces and returned to the atmosphere. The cooling ambient air temperature quickly approaches the dew point temperature, or the temperature to which the air must be cooled in order for it to become saturated, assuming pressure and moisture content remain constant. As the air temperature over the grass and deck railings surfaces dips below the freezing mark approaching the dew point temperature, often at daybreak, some of the air's moisture converts from the gas state (water vapor) to the solid state (ice), depositing the ice crystals on the subfreezing surfaces, allowing for the morning frost to spread across the landscape.

A morning's blanket of frost will disappear quickly with the warming rays from the rising sun; nevertheless, the taste and smells that surround that morning's first cup of coffee will always linger long after the frost disappears.

Heat Wave Not Alarming Episode

While first coming to South Dakota in the late '90s, I left Texas during one of its hottest summers ever recorded. My dog and I drove 1,200 miles north (I drove most of the way, he complained) anticipating relief from the scorching heat in the cooler climes of the Black Hills. At about 1,150 miles into the trip, I began to question my investment of time and money in getting a master's degree in meteorology. Along Highway 79, somewhere between Hot Springs and Rapid City, the temperature reached 105 degrees as the car's freon pressure waned. My dog, Wally, suddenly forgot all that stuff about being man's best friend, by expressing his disapproval with nonstop panting and sticking his tongue out in my direction — way out.

A slow retreating ice glacier (Yahoo Images)

At that time, June, July and August 1998, was the hottest three-month period in U.S. history, with the Northern Plains being no exception. The South Dakota state capital, Pierre, had the nation's high temperature on September 10, a dubious honor customarily reserved for places like Yuma, Laredo, Thermal and Death Valley. The "South" in South Dakota distinguishes us from that other Dakota to the north, and most certainly not implying a hot, muggy southern climate.

What gives with the abnormal heat? Is our climate changing? Is the world heating up along with Highway 79, causing glaciers to melt away, raising the sea level, and changing the landscape of the shorelines forever?

The simple answer is: Yes, of course, the climate is changing. In fact, the climate has been evolving ever since the Big Bang. Not too many thousand years ago immense glaciers covered the Midwest, pushing southward, forming the vast flattened expanse of the Great Plains. We can conclude climate warming continues today based on past climate records documenting increased glacier melt, allowing the current glaciers to continue their rapid retreat northward. Given those record-breaking hot temperatures back in the late '90s, it's difficult to believe the warming trend will end anytime soon. Back then, ask anyone in Pierre, or my auto air conditioning repairman, or my dog for that matter, and they will tell you the climate, along with their disposition, is heating up.

As climate changes, plants and animals either evolve with the changing climate, or they perish. Check out the remains of all those extinct animals now residing in the geological museum at the South Dakota School of Mines and Technology, if you have any doubt.

Meaningful climate change happens typically over geologic time scales, on the order of tens of thousands of years, not over decades or even centuries. Unfortunately, we perceive climate change over our lifetime. As we experience weather abnormalities, we think the climate must be changing and changing rapidly. We often extrapolate these short-term weather abnormalities into extreme weather changes a decade or even a century down the road, possibly concluding that based on the past few summers, our future summers here in the Black Hills will be more like Death Valley.

As we live our lives on this planet, we participate in the warming process, whether we like it or not. Scientific studies over the past 30 years conclude that man continues to speed up the warming process.

However, despite our ever-expanding influences, I believe Mother Nature retains some control of her realm with checks and balances. For example, if indeed our climate is warming rapidly, and the melting glaciers are spreading more water across the globe, the warmer and moister environment will lead to a more unstable atmosphere. An unstable atmosphere, in turn, will lead to more clouds, reducing the incoming solar radiation to at least slow the progression of climate warming.

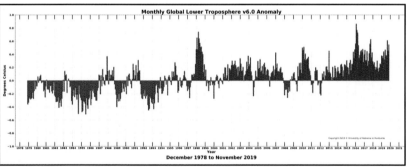

Monthly global near-surface temperature anomaly showing more frequently warmer global temperatures beginninng in the late 1990's — December 1978 through April 2019 (University of Alabama in Huntsville)

I believe the ever more prevalent, above-normal heat we have endured across South Dakota, the country and around the globe, should be viewed as something not more than a hiccup in this century's weather occurrences. In the relatively short-term weather happenings, keep in mind that weather is not climate. With that said, in the past few decades of out of the ordinary rapid global warming, coupled with mankind's growing impact on the environment, the long-term abnormal heat appears to be the new normal.

Raindrops Keep Falling on My Roof and Forecast

Since most people soon forget those uneventful weather events that occurred a few weeks back, no less a few years back, I will expand on the weather amnesia point first, before discussing one of my memorable forecast lessons.

We can attribute the good old "out of sight, out of mind" syndrome to the simple fact that most people's weather interest centers on current and future conditions for planning (or, in my case, lack of preparation). For example, because last week's rain event holds little value to us as we plan for today's and possibly tomorrow's activities, we tend to forget about it. Enhanced value exists in knowing something about tomorrow's rain or next week's rain event.

To illustrate from personal experience: my roof leaked during one heavy rain event some years ago. So, I needed to know when it will rain on my rooftop again and, more specifically, into the hole that only rainwater can locate. Knowing that the predicted weather conditions can give me a timeline to schedule roof repairs — the forecast has value. If the forecast calls for sunny and dry weather, it makes no sense for me to fix the roof when I could be out enjoying the sunshine. After all, how can you find the leak when it's not raining. If the forecast calls for some probability of rain, I play the odds that the raindrops will fall on someone else's watertight roof and avoid mine. Still, I will use that information to place pots at the ready.

A countless number of raindrops keep falling on this roof (Google Images)

If I lose this game of chance, I can quickly set out the pots to catch the leak while rationalizing that climbing around on a wet and slippery roof would be foolish. For me, fixing a leaky roof can be a long-drawn-out process, but there still remains some value in knowing tomorrow's forecast for rain. This same principle applies to replacing badly worn windshield wipers.

So, now that we realize why we are mostly forward-thinking people, a thought-provoking but common weather phenomenon that played havoc with my forecast happened to me when I first started forecasting in the U.S. Air Force. As discussed earlier, like most people I tend to forget those humdrum missed forecasts in the past, but not this forecast due to the fundamental lesson I learned about forecasting the weather, mainly when a three-star general depended on it to be right.

A cold front moved through Barksdale Air Force Base, LA bringing cloudy skies, cooler temperatures along with brisk northerly winds. My forecast for the following day called for continued cloudy skies, with the winds shifting from a northerly direction to southeasterly, suggesting the cold air would not penetrate too far southward and begin to retreat to the north. I also predicted light rain and a high temperature in the mid-50s or about 10 degrees warmer than the previous day's high temperature. I nailed that forecast with one exception, the temperature, which turned out to be 10 degrees cooler than the previous day's high temperature.

Why would the high temperature for the following day be much cooler than the previous day's high temperature? After all, the sky conditions were the same, and those southerly winds mean warmer temperatures.

The answer to my missed forecast lies with the transfer of energy from the air to the raindrop. When transferring energy from one body to the next, the body losing the energy loses heat and cools. This dynamic interplay between suppliers and consumers of energy, on either small or large scales, causes the weather to change from moment to moment, day to day or over even lengthier periods.

Typically, here in South Dakota, the passage of a cold front means the arrival of less humid Canadian air. When cooler raindrops, relative to the surrounding Canadian air, mix into the drier air, the energy transfer from the dry and relatively warmer air to the cooler raindrop begins. The sequence goes like this: the relatively warmer air provides energy to the raindrop, the raindrop uses the energy to evaporate partially, and therefore the loss of energy/heat from the air cools the air temperature. In dry and hot climates, evaporative coolers use this same principle to cool homes and businesses effectively.

So, there I was, a nervous first lieutenant trying to explain this previous day's temperature forecast snafu to a three-star general. The lesson burned permanently in my memory bank from that day forward; raindrops cool the air, and at times, they cool the air so much that the rain may even change into snow.

Shedding Light on Wind's Chilling Effect

On those blustery days with strong winds and the thermometer registering temperatures near freezing, the term wind chill appears more frequently in print, on your smartphone, on radio, and on television. The wind chill suggests unsafe temperatures well below freezing, capable of causing frostbite. Probably more than any other weather term, wind chill creates a great deal of confusion.

Let me explain just what the term wind chill means, by first stating what it is not. The wind chill is not a measure of the ambient air temperature. If the outside thermometer at the airport reads 40 degrees Fahrenheit and the wind chill index registers a below freezing 20 degrees Fahrenheit, the temperature of the air, whether moving or not, remains 40 degrees Fahrenheit. Even with the ambient air temperature near freezing, but the wind chill reading is well below freezing, you can still place your tongue on the pump handle without any fear of it permanently sticking.

High winds carry a lot of force, but never enough to repeal the second law of thermodynamics, which says that heat cannot flow from a cold object to a warmer object. In more practical terms, this means that if the outside air temperature is 40 degrees Fahrenheit, any outdoor non-living object, like an automobile engine or a pump handle, cannot cool down below 40 degrees Fahrenheit. You see, for any outdoor physical body to cool below the outside ambient air temperature that object must give off heat to the warmer outside air. No way says the second law of thermodynamics.

When we throw into the mix an accompanying cooling process along with your human body, your skin temperature can indeed drop below freezing on those windy days. Automobile engines or pump handles do not give off moisture, i.e., sweat, but your body does perspire through its skin. As the moisture passing through your skin evaporates into the air, it removes the heat from the air near the skin's surface to evaporate the perspiration. This evaporative cooling can indeed cool the skin well below the air temperature to the much talked about wind chill index. On those days with near freezing air temperatures and strong winds that can quickly remove the layer of air near your skin, allowing the evaporation cooling process to accelerate, frostbite on exposed skin can develop swiftly.

So, if the actual temperature is above freezing, your wet tongue will feel colder than the air temperature, but no way will it establish a semi-permanent frosty union with the pump handle, even if a 20-mile per hour wind drops the wind chill to a below freezing index.

Wind Chill Genesis

The term wind chill goes back to the Antarctic explorer Paul A. Siple, and Charles Passel who coined the term in 1939, based on experiments to study the time needed to freeze water in plastic cylinders exposed to the elements. They found that the freezing time depends on the temperature of the water at the start of the experiment, the outside temperature and wind speed. Keep in mind that the higher the wind speed, the more efficient the evaporation process at the water's surface in the plastic cylinder, resulting in a faster time to freeze the water's surface.

The old wind chill index, used by the National Weather Service (NWS) since 1973, more than likely over-stated how cold the air feels on one's exposed skin. After more than a decade of criticism from scientists, in 2001, the winds of change blew through the way the NWS calculated the wind chill. After much testing, discussion and coordination among various weather groups in the U.S. and Canada, an agreement was reached on the new formula. So, out with the old and in with the new way of calculating the effects of heat loss on our skin due to cold.

Cooling power of wind expressed as an equivalent chill temperature (under calm conditions)												
ESTIMATED WIND SPEED (IN MPH)	**ACTUAL THERMOMETER READING (F)**											
	50	40	30	20	10	0	-10	-20	-30	-40	-50	-60
	EQUIVALENT TEMPERATURES (F)											
Calm	50	40	30	20	10	0	-10	-20	-30	-40	-50	-60
5	48	37	27	16	6	-5	-15	-26	-36	-47	-57	-68
10	40	28	16	4	-9	-24	-33	-46	-58	-70	-83	-95
15	36	22	9	-5	-18	-32	-45	-58	-72	-85	-99	-112
20	32	18	4	-10	-25	-39	-53	-67	-82	-96	-110	-124
25	30	16	0	-15	-29	-44	-59	-74	-88	-104	-118	-133
30	28	13	-2	-18	-33	-48	-63	-79	-94	-109	-125	-140
35	27	11	-4	-21	-35	-51	-67	-82	-98	-113	-129	-145
40	26	10	-6	-21	-37	-53	-69	-85	-100	-116	-132	-148
Winds greater than 40 MPH have little additional effect.	LITTLE DANGER (for properly clothed person) Maximum danger of false sense of security.			INCREASING DANGER Danger from freezing of exposed flesh.			GREAT DANGER					
	Trench foot and immersion foot may occur at any point on this chart.											

Monthly global near-surface temperature anomaly showing more frequently warmer global temperatures beginning in the late 1990's — December 1978 through April 2019 (University of Alabama in Huntsville)

The current wind-chill formula measures temperatures closer to the ground, at about the height of a human face, where the wind chill can cause the most harm. Also, actual people tested the current formula in wind tunnels and climatic chambers. This updated formula remains the one we still use today, with more-realistic values on how wind and cold can affect our bodies, but the ambient air temperature remains unchanged.

Susan Sanders, the warning coordination meteorologist with the Rapid City NWS, states that the new is better than the old in three specific ways. "It incorporates heat transfer from the human body. Obviously, a person keeps producing heat to keep warm, unlike water and ice that eventually will cool to the surrounding air temperature," she said. "(It) Reduces the wind speed to a value more representative at the height of the average face (five feet above the ground). We measure winds at 10 meters above the ground; this index uses a reduction factor to account for greater friction near the ground. "The new index also correlates the time it takes to develop frostbite to specific wind-chill values. This helped us establish our new warning and advisory criteria."

Susan goes on to say that the Rapid City NWS office will use 40 degrees Fahrenheit below zero for wind-chill warnings and 25 to 40 degrees Fahrenheit below zero for wind-chill advisories. So, listen to Susan, and pay attention to all advisories and warnings, because the current wind-chill warning and advisory criteria, coupled with the current calculation of how long it will take exposed skin to develop frostbite, indeed can save your skin, if not your life.

It's important to note that the wind chill index is only a subjective approximation, since how one feels the chill depends on not only temperature and wind speed, but the amount of exposed flesh, the type of clothing worn, the amount of sunshine, the relative humidity and the physical condition of the person.

When Cold Means Cold

The record cold for Rapid City is 34 degrees Fahrenheit below zero reported on February 11, 1899. The South Dakota record cold is 57 degrees Fahrenheit below zero recorded in McIntosh on February 17, 1936. Considering the high frequency of back-to-back-to-back mild winters often enjoyed in the Black Hills in the past, one can lose perspective as to what constitutes a frigid-cold winter day versus a cold winter day or even a mild winter day.

No clear line in the snow separates frigid-cold from just plain cold, or for that matter, just plain cold from mild. Frigid-cold is a relative term but I, for one, know a frigid-cold winter day from a cold winter day. The word "frigid" has nothing to do with some number you see flash across a bank marque that represents a temperature. Given certain conditions, some 30-degree Fahrenheit days feel like 60-degree Fahrenheit days, yet, other 30-degree Fahrenheit days that engages all the bodily senses, and not just the sense of touch, are not just cold days, but frigid-cold days, end of discussion.

Indeed, it doesn't take a meteorologist to know that frigid-cold vigorously attacks the sense of touch, like placing one's tongue on the pump handle, for example. Nevertheless, unadulterated frigid-cold temperatures also invade the sense of sight. The sight of steam fog billowing off the creeks and crystallizing its moisture as feathery decorations on the nearby grass, leaves, and trees helps the brain convince the body that, "Baby it's not just cold outside but frigid-cold outside."

A frigid and sunlit winter day in the Black Hills, SD (Teresa Riggio)

The sense of hearing adds its own import to the transition from a cold winter day to a frigid-cold winter day. On a frigid-cold day, the slightest of sounds, such as a distant bird chirp or a step in the snow, is more easily detected. Frigid-cold somehow dampens the surrounding "white" noise created by the daily hustle and bustle activities. When it's frigid-cold outside, it appears to the eye that moving objects seem to slow down somehow, reducing their noise level, allowing for a ghostly silence to spread across the landscape. So very quiet, one can hear the faintest vibrations on a frigid-cold day.

And, of course, taste and smell — no one can tell me that you cannot taste and smell frigid-cold. At least in my memory bank, frigid cold air tastes and smells clean and fresh. That distinctive taste and smell in the air on a frigid-cold freezing day reminds me of many cold winter days playing as a child on the streets of Chicago. I cannot explain the taste and smell of frigid-cold just as I cannot describe a color. The taste and smell of frigid-cold compare well to the taste and smell of pure water. Both are entirely refreshing and quenching, even without experiencing either the taste or smell.

Dan Lutz

Barbwire fences and prairie grass do very little to slow down the wind. When wind enters the harsh weather mix, further misery often raises its ugly head.

🍂 THE WIND WE SENSE

It's almost a national pastime to talk about, and on many occasions, complain about the weather. I'm convinced most of the complaints have little to do with too cold, too warm, too dry or too wet. Strong blowing and howling wind cover the preponderance of grievances. When wind enters the harsh weather mix, further misery often raises its ugly head. Wind with frigid temperatures means bone-chilling wind chill. Wind with snow offers up blinding blizzards. Dry wind suggests choking dust storms, drought and sinus headaches. Wind with high humidity and dark skies often indicates severe thunderstorms and property damage.

I believe that the public's complaints about the weather would be reduced drastically by merely taking the wind out of the equation. That would be impossible to do here in the Northern Plains, where very few wind barriers stand in the way of those strong winds blowing south out of Canada. Barbwire fences and prairie grass do very little to slow down the wind.

Jelly Sandwich Explains Wind

In the 1960s protest anthem, "Subterranean Homesick Blues," Bob Dylan wrote, "You don't need a weatherman to tell you which way the wind blows." And indeed, if you want to know current wind direction, moving leaves, ripples on a lake, flapping flags or the wet finger in the air will work just fine.

But, if you want to know the wind direction tomorrow or someplace other than your current location, look at your local paper's weather map. The wind always blows from high-pressure areas (the big blue "H" depicted on weather maps) toward low-pressure areas (the big red "L" on weather maps).

If you want a better understanding of this fundamental law of nature, you can perform a simple experiment in the comfort of your kitchen. Take two slices of bread, some jelly, and a knife. Spread the jelly generously over the bread, making a jelly sandwich. You may now recruit any child around the age of six to eat the sandwich.

Assume for the moment that the child's hands holding the sandwich represent the big blue "H" or high pressure. Taking this lunchtime weather experiment one step further, the jelly between the bread slices represents the wind. To get a firm grip of the sandwich the child's hands apply pressure to its center, the jelly (now representing wind) oozes away from

The squeezing-the-jelly-sandwich experiment (Dan Lutz)

the center of the jelly sandwich to areas of less pressure, and more than likely ends up on said child's face and clothes.

Like the jelly in the sandwich, the wind always blows away from high pressure toward lower pressure. Now you ask: If the wind blows from a high-pressure area toward a low-pressure area, why is it that weather maps and the weatherperson on TV indicate winds blow clockwise around high-pressure systems and counterclockwise around low-pressure systems?

This wind-related phenomenon considers life on a spinning and round plant, which adds some complexity to our simple jelly sandwich law. Remember our old friend Coriolis force? Let me try to enlighten the reader a little more about this bizarre force.

If you were observing the wind located on this spinning planet from far out in outer space, the observed wind would indeed directly blow from high-pressure regions straight toward low-pressure areas, confirming the law of the jelly sandwich. Fortunately, we observe and sense the wind here on this round ball we call Earth, which spins as fast as 1,036 miles per hour. The spinning of this round ball causes the wind to appear to move in a curved motion to us standing on Earth and not in a straight line. In the Northern Hemisphere, a straight-line wind will appear (to spinning Earth-bound residents) to bend to the right. Therefore, the straight-line wind blowing from an area of high pressure will turn to the right giving it a clockwise rotation. Conversely, the wind blowing into a low-pressure area also will bend to the right giving it the accustomed counterclockwise motion.

All bets are off if you find yourself in the Southern Hemisphere, where Coriolis force reverses itself causing a straight-line wind to appear (to spinning Earth-bound residents) to bend to the left. Therefore, the straight-line wind from an area of high pressure will turn to the left, giving it a counterclockwise rotation. Conversely, the wind blowing into a low-pressure area also will bend to the left, giving it the familiar clockwise motion to those folks living down below in Australia.

Our old friend, the Coriolis force, causes the apparent right turning force in the Northern Hemisphere and the left turning force in the Southern Hemisphere.

Jelly Sandwich Law in the Third Dimension

To review the jelly sandwich law: when pressing on the center of a jelly sandwich (representing a high-pressure area) the jelly (representing the wind) tends to squirt outward at the ends of the sandwich in all directions to areas of less pressure. This tasty, albeit messy experiment, symbolizes why the wind blows from high-pressure areas to low-pressure areas.

Now, there may be a few thoughtful weather aficionados out there who, on reflection, realize that yours truly presented only two dimensions of the three-dimensional story, the north/south, and east/west winds, but failed to discuss the third wind direction, which is equally as important. You see, for the Jelly Sandwich Law to remain intact, there must be a wind blowing up.

Remember, as discussed earlier, the air that we work in, play in and, breathe into our lungs has weight. Also, recall that pressure measures the weight of the air. Therefore, the more weight overhead, the greater the pressure at any one location.

Now it stands to reason that in the column of air molecules stacked on top of each other, the more substantial part of the air column, or highest pressure, is located at the Earth's surface. The higher the elevation, the lighter the air or, the less the air pressure, because there would be fewer molecules stacked on top of one another.

So, now we are living in a world of higher pressure at the Earth's surface relative to the air pressure aloft. Of interest, at about 22,000 feet altitude, the air is at its central vertical compression, the air above that altitude will weigh the same as the air below. Now going back to the "Jelly Sandwich Law," if the theory holds, there must be a wind blowing from the surface (higher pressure) upward into the heavens (lower pressure). Otherwise, the "Jelly Sandwich Law" crumbles.

We know that while walking around the Black Hills or anywhere outside, we have no concern about our hair standing on end due to some wind blowing upwards from the ground. Why not you may ask. Because the force of good old gravity keeps the vertical wind in check. As the "Jelly Sandwich Law" works its magic to push the surface air skyward, gravity does the opposite to hold it back.

This up/down balancing act by nature is known as the atmospheric hydrostatic equilibrium. The hydrostatic equilibrium keeps the weight of our atmosphere from entirely slipping down on us and yet provides some holding power to prevent it from escaping out into space.

So, there you have it. The "Jelly Sandwich Law" holds together no matter which way the wind blows. Adding a little peanut butter to the jelly sandwich also makes for a delicate balance between slipping down your throat and sticking to the roof of your mouth.

Weather, Thar' She Blows

So why all the strong winds here in the Northern Plains? We can blame those cold, dense and heavy high-pressure systems that drop out of Canada and slide southward along the Rocky Mountain Front Range. Wind speed depends on the difference in pressure between the areas of high pressure and low pressure. The higher the pressure difference, the greater the wind speed. By squeezing the center of the jelly sandwich with higher pressure, the jelly squirts out the side with increased velocity. By increasing the pressure on the open end of a garden hose, the water squirts out with more speed.

During the winter season, these cold and dense domes of air pressure sliding out of Canada and over the Northern Plains exert immense amounts of pressure on the air near the ground. The increased pressure, in turn, generates strong winds as the air squeezes away from the center of these large high-pressure systems. The more substantial the weight of the high-pressure system, the higher the pressure difference, and the faster the winds blow. Only when the pressure gradient weakens as the center of the high-pressure system either moves off or warms to become less dense, will the winds begin to subside.

Year round, the Northern Plains typically experience stronger winds than all other areas in the U.S., because the coldest and most dense domes of high-pressure track from the Arctic Plains across the Northern Plains and continue southward into the Dakotas. Among those states making up the Great Plains, North Dakota ranks the highest with the strongest winds, while the Southern state's rate lower. As these cold and dense air masses modify, becoming warmer and less dense on their journey southward, the pressure gradient weakens along with wind strength.

Not surprisingly, maximum warming of these once dense and cold air masses occurs by the time they slide into the Southeastern regions of the U.S. As one would expect, it's the Southeastern regions of the country where the least year-round wind occurs.

So, don't blame the weatherman for the strong winds. Blame our neighbors to the north.

Appreciate the Northwestern Wind

Autumn's daily temperature patterns typically follow short-term mood swings, with warm days followed by chilly days only to relish in the return of warm days. This up-down ride of temperature, with time, will become more down than up as the days turn shorter and colder with the calendars' slow march to winter. Also, with the approach of winter, the back-and-forth fetch of wind direction from southerly to northerly and back to southerly will settle in favor of a more stable northerly direction.

As discussed above, large high-pressure areas of dense cold air build over the colder climes of Northern Canada as the summer slowly fades into autumn. As these cold air masses become more substantial, denser, and heavier with increasing cold, good old gravity and a push from the Polar Jet Stream sends them southward across the Northern Plains. The sheer weight of these massive areas of high pressure spreads the cold air southward, producing gale-force winds wailing over the Dakota prairies, a typical winter occurrence.

From the perspective of the Northern Plains, the preponderance of moving air flows from the northwest toward the southeast. Assuredly as the northern Canadian temperatures fall as winter approaches, the nippy northwest flow will increase, ushering in a growing number of colder autumn and winter days. A northwesterly wind — or some slight variation — endures not only during the autumn and winter but all year long. It is the predominant wind pattern across the Northern Plains.

Thankfully, even during the summer's heat, the more familiar wind in the Northern Plains originates from the northwest. Despite the record books documenting record-breaking hot and humid summer days under the influence of the more tropical southeast wind pattern, gratefully the northwest wind prevails, even during the summer months, providing occasional relief from excessive heat.

This overabundance of breezes from the northwest provides significant comfort to our year-round outdoor activities. The northwest wind pattern, coupled with the height of the Rocky Mountains, the Big Horns or the Black Hills, takes the bite out of winter's cold and our summer's heat and humidity. During many wintry days, those regions on the leeward side of the mountains and hills find themselves in the shadow of the northwest winds and welcome the warmth created by the chinook winds.

The northwest winds carry the air over the mountains and hills, and down their backside, sinking into the eastern foothills and prairies below. The sinking air travels from lower pressure aloft, say near Lead/Deadwood in the Black Hills, to the higher pressure below, say Rapid City, squeezing the air molecules closer together,

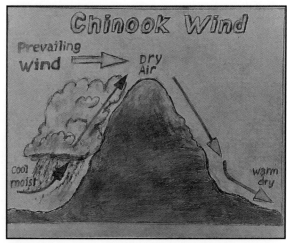

The warming process of the downslope chinook winds (Meg Roman)

which in turn gives off energy/heat, heretofore energy needed to keep the air molecules apart. The added heat to the environment can do nothing else but warm the air at the lower elevations, providing much appreciated relief from the winter's biting cold.

The temperature change described here, due primarily to the chinook winds, set a world record for the fastest temperature change which still stands today. On

January 22, 1943, Spearfish, SD, set and still holds the record for fastest temperature change. That day started out very cold in Spearfish with a 7:30 AM MST temperature recorded at -4 degrees Fahrenheit, and two minutes later at 7:32 AM, the temperature rose 49 degrees Fahrenheit to +45 degress Fahrenheit.

Here in the Black Hills, a breeze of any magnitude from the northwest offers significant comfort to our year-round outdoor activity. We enjoy many days of northwest winds throughout the year. The summertime northwesterly wind pattern brings not only more cooling from Canada but also less humid air, allowing for even more cooling of the Earth's surface after sunset. This after-sunset cooling effect makes for more comfortable evening conditions after a hot summer afternoon.

The wintertime northwesterly winds blowing across the Black Hills bring warming relief to the prairies below on the tail of those chinook winds. Warming, snow-melting temperatures can quickly follow a few days of blizzard-like weather erasing all memories of those white-out surroundings and subzero wind chill values.

Bottom line is, regardless of the occasional blustery autumn and winter days, expect more than our share of mild days thanks to the downward flow of air from the top of the Black Hills to the lower elevations below, often spreading as far east as the Missouri River. Also, the downward airflow along the eastern slopes of the Black Hills dries the air, allowing for the summer's afternoon heat to more readily escape into the heavens, comfortably cooling those midsummer evenings.

Northwest Winds Dominate in the Northern Plains

The occasionally unexpected turn of the elements often brings out the worst of the weather beast. Despite this proclamation of the atmosphere, we can take comfort in the knowledge that there will be some certainty with a few weather-related encounters.

With the arrival of fall and the subsequent downward slide of temperatures across the Northern Plains, with great certainty, you can count on the turning of the leaves, pumpkin pie, the opening of pheasant season, warm and cozy sweaters, football games and my all-time favorite, Thanksgiving. All of these fall events are sure to be touched by the weather.

A freshening northwest surface wind remains yet another absolute occurrence this fall. Who can forget those crisp autumn days highlighted by that brisk northwest wind fanning the newly fallen leaves across the landscape? Toward the end of the summer-to-winter transition, the days occupied by the nippy northwest flow will increase in number, ushering in a growing sum of colder days.

Those regions finding themselves in the shadow of the northwest winds as they cross over the mountains and hills create the chinook winds, as described previously, so vital to creating the many mild days we enjoy yearlong living here in the shadow of the Black Hills. The sinking air traveling from lower pressure aloft to higher pressure below operates on the same principle as a pressure cooker.

According to the gas law (pressure = volume x temperature), by keeping volume the same, increasing pressure increases temperature. A pressure cooker boils a liquid (usually water), releasing steam within a confined space which in turn increases the pressure inside the cooker, allowing the temperature to rise. This principle enables the pressure cooker to cook foods faster than conventional methods. You can feel an increase in heat in the palm of your hand just by doing the simple act of creating a tight fist to increase the pressure.

So, Bob Dylan was partially correct when he sang back in 1967, "You don't need a weatherman to know which way the wind blows." Across the Northern Plains, an educated guess would be from the northwest, and you would be correct many more times than not.

The Black Hills Push Wind Speed

The region surrounding the Black Hills most assuredly gets its fair share of unusual weather that covers the entire weather spectrum — from strong thunderstorm-force downburst winds to hurricane-strength straight-line winds to blizzards producing copious amounts of heavy wet snow to unbelievable warming and cooling over short periods.

One year, 18 inches of snow fell at Rapid City Regional Airport, setting a new 24-hour snowfall record. In Spearfish, temperatures climbed 49 degrees Fahrenheit from 4 degrees Fahrenheit below zero to 45 degrees Fahrenheit above zero in about 150 beats of the heart. Crazy weather, yes, but not atypical here in the Black Hills. Not all extreme weather — but much of it — happens in this region of South Dakota because of one highly conspicuous wild and wacky weather-maker.

This crazed weather-maker is not solely about some combination of a high-pressure system, a low-pressure area, a jet stream, or a cold front. In fact, it has nothing to do with anything moving around in our gaseous atmosphere. Our extravagant weather-maker has everything to do with this area's rock-solid Earth-bound geologic landscape.

Those beautiful Black Hills can and do ratchet up much of our weather patterns into the severe category. One instance of the Black Hills influence was a high-wind event when non-thunderstorm wind gusts reached 86 miles per hour at Ellsworth Air Force Base, 71 miles per hour at Tilford and 65 miles per hour just north of Hereford in Meade County.

Those unfamiliar might think surface winds rarely reach these hurricane-like speeds. Not so. Those atmospheric weather patterns that often produce strong winds generally occur once or twice a year. A high-pressure system to the west squeezed by an intense low-pressure system sliding southeast from Western North Dakota past the northeastern foothills of the Black Hills toward southeast South Dakota often results in high winds. By including the Black Hills in this atmospheric equation, high winds can and do become extreme winds.

It's not out of the question that the influence of the Black Hills adds 20 to 25 miles per hour to wind speeds. The Black Hills, in effect, enhance the high-pressure system to the west, thus adding substantially to the pressure gradient, and finally increasing the wind speed.

Let me explain. We can agree that the speed with which a skier moves downhill depends mostly on the slope of the mountain. The steeper the slope, the faster the skier travels down the hill. Picture a skier following the plumb line down the gradual incline of the beginner slope. That skier will not go too fast. Now place that same skier on the much steeper triple black-diamond slope, and Katy-bar-the-door; the skier reaches the bottom so much sooner to get back in the chairlift line.

The same principle applies to the high-wind events in and around the Black Hills. High pressure to the west, coupled with a passing low pressure system to the east, produces a black-diamond steep slope of decreasing air pressure over a short distance. Now, by introducing the Black Hills into the mix, the physical presence of the Black Hills amplifies the high-pressure area to the west, further squeezing the winds against the eastern slopes of the Black Hills, and lo and behold the black-diamond hill becomes a triple black-diamond slope, forcing the winds to slide down this high-pressure slope at incredible speeds. Winds that at times can reach hurricane speeds over 73 miles per hour.

Upper-Level Low Makes for South Dakota Wind

Get this: According to the American Energy Association, South Dakota ranks as the fourth-windiest state in the union behind top-ranked North Dakota, second-ranked Texas, and Kansas. I'll concede North Dakota's top spot, but South Dakota should have a lock on second place.

My tongue-in-cheek reasoning is this: the highspeed prevailing wind in the Northern Plains blows out of the northwest across South Dakota and into Nebraska and Kansas and so on. Now, there might not be a whole lot of trees, hills, fences or buildings between South Dakota and Kansas, but surely there must be enough obstacles, such as countless acres of corn, to slow that northwest wind a tad before it reaches Dorothy's farmhouse in Kansas. Therefore, I'll place South Dakota ahead of Kansas on the windiest-state list.

I lived in the Texas wind. I know Texas wind and Texas does not have the high wind speeds we experience here in South Dakota. Move South Dakota ahead of Texas.

North Dakota can stay in the No. 1 slot because the northerly winds probably slow a bit traveling across the vast prairies to the north before crossing into South Dakota. To carry this north-winds slower-moving-south thought process a step or two further, Montana should be third, instead of fifth as listed by the American Energy Association, placing Wyoming fourth instead of seventh.

If those energy folks lived here in the Black Hills, I know they would come around to my way of ordering the top four windiest states. It's not uncommon to experience afternoon winds around Rapid City persisting above 30 miles per hour with gusts exceeding 40 miles per hour for relatively long periods.

Interestingly, the hourly wind pattern on those blustery occasions appears similar from day to day. The winds increase quickly at 8 AM to a secondary peak about 11 AM, then decrease slightly just after the noon hour, only to intensify again to the day's peak level by mid-afternoon. Finally, after sunset, the winds would ebb.

The multi-day extended longevity of these high wind events typically results from a broad and deep upper-level low embedded in the jet stream and located high above the Great Lakes region. Typically, upper-level lows move along quickly by the jet stream, impacting any one place for a brief period before moving on toward the east.

Detached upper-level cutoff low over the Great Lakes region

Not so on these windy multi-day cycles, when the upper-level low becomes detached from the jet stream's grip. Like a spinning top on a well-balanced pool table, it remains fixed over the Great Lakes for extended periods. The backside winds of this upper-level low rush southward across the Northern Plains with tremendous speed and persistence.

During the day, afternoon heating from the sun mixes those strong overhead winds with the surface winds, giving rise to strong and gusty surface winds, especially when that upper-level low remains fixed. At night the air near the ground cools rapidly, and the mixing of the atmosphere subsides with loss of solar heating, causing the stronger upper-level winds to decouple from the surface winds, allowing the surface winds to lessen in intensity.

Like that spinning top, with time, the upper low begins to lose its energy and staying power, and the multiple days of well above average wind speeds start to retreat to normalcy.

Jet Stream Drives Winter Weather

A time of notable weather transition begins during both the autumnal and spring equinox when the sun finds itself directly over the equator and night and day are approximately the same length everywhere across the globe. During the autumnal equinox time of change, periods of cold days become intersperse among periods of hot days with increasing gusto as the sun journeys southward and the Northern Hemisphere's winter season approaches. During the spring equinox time of change, periods of warmer days become intersperse among periods of cold days with increasing enthusiasm as the sun now journeys northward and the Northern Hemisphere's summer season approaches.

One essential element of these large-scale seasonal changes (summer to winter and back again) finds itself within the broad ribbon of westerly winds that relentlessly blow high in the atmosphere from a generally westerly direction to the east, across North America and around the globe. These embedded winds, with speeds two to three times faster than its surrounding westerly current of air, make up the polar jet stream. The polar jet stream with its many loops, branches, and changing direction and altitude, serves to move large air pressure masses across North America, including those cold high-pressure systems that evolve over the North American Arctic region.

The very presence of a robust jet stream high in the atmosphere sets in motion some of the most intense snowstorms that rage across the Great Plains. Wind speeds in the core of the polar jet stream can easily exceed 150 to 200 miles per hour at an elevation of 25,000 to 30,000 feet. As the core of the polar jet stream surges northeastward from the southwest and into the Great Plains, it brings with it tons of mid-level moisture from the South Pacific and low-level Gulf moisture. This abundance of moisture rides northward, often resulting in intense blizzard-producing surface storms. These winter storms develop quickly along the lee side of the Rocky Mountains, tracking with the northeastern-flowing jet stream into the Upper Plains and Midwest. Some of our most intense winter storms first develop in the Four Corners region of the country and intensify as they move northeastward, directed by the winter jet stream highway.

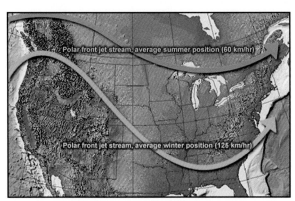

The summer and winter polar jet stream locations over North America (Yahoo Images)

The polar jet stream, the main highway of winter storms, not only aids in the development of these winter storms, it foretells the direction and speed of movement of these winter storms and the duration of frigid temperatures.

During the extremely cold winter season, when frigid, dry Arctic air spills southward out of Canada along the front range and often well into the Gulf of Mexico, you will find the polar jet stream dropping far south of the Northern Plains into the Southern Plains. Conversely, during periods of mild winter weather and the summer months, the jet stream is well established over lower Canada.

Pilots ferrying aircraft to Britain from the United States during World War II first discovered the polar jet stream. They encountered much faster times than expected crossing the Atlantic in the winter months, with tailwinds of up to 150 miles per hour at higher altitudes. Today, airlines regularly rout eastward-bound passenger aircraft into this polar jet stream and westward bound passenger jets away from the polar jet stream to save time and fuel.

Whether you are concerned about saving jet fuel or adding fuel to the furnace in anticipation of those cold winter nights, the polar jet stream remains the foremost topic of weather discussion during the winter season, as well as an excellent indicator of the severity of the winter season.

Wind Takes on Many Different Roles

Early spring winds seem to do their job well. Case in point: one beautiful spring day while walking my dog Wally, I noticed many kites of all shapes, sizes, color and length of tail dancing across the blue sky thanks to those invisible streams of moving air. Only a thin string held tightly by a group of ageless kids kept those kites safely away from the always present and dreaded kite-eating trees. Unfortunately, the law of averages dictates that, on occasion, a wayward gust of wind will eventually provide one of those trees a tasty treat.

Distinct gradients of air pressure cause the wind to blow across the prairies, through the pines, and over the parks, allowing those kites to float high in the sky. The greatest contrast of air pressure lies between the centers of the high-pressure and low-pressure systems. Those northwest winds, prevalent to the Northern Plains, tend to pick up speed markedly when found smack dab between a departing low-pressure area and the soon-to-arrive high-pressure area.

A boy flying a kite (Yahoo Images)

The second law of thermodynamics tells us that nature is uneasy with dissension among the atmospheric ranks. Nothing alarms nature more than low-pressure areas next to high-pressure areas, dry bumping into humid, or cold overtaking hot. Nature uses the wind at all levels to signal its effort to regulate the often highly dynamic atmosphere. The higher the differences between pressure, temperature, or humidity, the stronger the wind blows, as nature restores balance in the atmosphere.

When restoring balance to our atmosphere, winds signal the arrival of good weather and not-so-good weather. The winds indicate masses of air, with different meteorological characteristics, coming together, often with an impact that promptly brings winter's blizzard-like conditions or spring's thunderstorm-like conditions across the Upper Plains. On many occasions, these same winds foretell the arrival of frigid cold from the north, followed by a gradual shift to the southwest transporting warm air to comfort our chilled bodies.

On a more practical level, in the spring the winds deposit seeds and distribute pollen across the prairies and hills to help replenish the landscape with trees, grasses, and flowers. During the summer, the evening winds, albeit calmer than their winter counterpart, can bring relief from the day's stifling heat.

On the downside, these same winds will pick up and push dust into our businesses, homes and, on a more personal note, into our respiratory system. These winds also will be a thief of water. The movement of the wind not only disturbs vast amounts of fertile topsoil but also contributes to the loss of immense quantities of surface water due to increased evaporation rates. Evaporation causes the removal of moisture in prairie plants, as well as water in our streams and lakes.

Interestingly, high winds, more than high temperatures, have the most significant impact on increased evaporation. And, evaporation rates influence crop production almost as much as rainfall. Loss of moisture in the prairie grasses leaves the area susceptible to fires. These same winds will not only help create the fuel for grass and other wildfires but also can push the flames across fields and into rural communities with lightning speed.

Alongside death and taxes, you can count on the wind to blow across the Northern Plains landscape on a daily basis. Unlike death and taxes, when the wind is at our doorstep, we can always literally and figuratively "go fly a kite."

A Windy Perspective from On Top

Some years ago, the National Park Service put on a spectacular firework display from high on top of Mount Rushmore. The cliché, "You've seen one, you've seen them all," does not hold true for that pyrotechnical exhibition.

The Mount Rushmore fireworks display will forever hold a special place in my psyche. During one memorable Fourth of July celebration, NewsCenter1 broadcast its morning, early evening and evening newscasts live from the base of the faces. Typically, during these remote newscasts, the weathercast comes from a location removed even farther from the already "remote" site.

The author firmly planted on top of Mount Rushmore, SD (Mark Simpson)

For this extravaganza, the early evening weathercast came from a granite plateau high on top of the stone head of George Washington. Thanks to the cooperation of Mount Rushmore personnel, our viewers received their televised weathercast from approximately 5,690 feet above mean sea level, and about 20 feet behind George Washington's forehead, where I firmly planted myself as if suction cups were attached to my bottom side.

Helicopters were used to get the fireworks to the top of the mountain. Unfortunately, NewsCenter1 personnel carried cameras, microwave equipment, monitors, and cables to the backside of George's head on our backs. After a 20-minute ascent up the rocky and steep mountainside, a 200-plus metal step staircase remained to be negotiated on shaking legs, burning lungs, and a pounding heart, to reach George's head at the summit.

Once on top, the equipment was set up to start the television signal's incredible journey from high on top of Mount Rushmore back to a satellite truck at its base, then up to a satellite located thousands of miles in outer space, and finally back down to the NewsCenter1 studio in Rapid City, SD. After much testing and calibrating, our early evening weathercast went across the airwaves from high atop of Mount Rushmore — an incredible feat back then from such a remote location.

From my lofty perspective, the day's afternoon winds were notably strong, the result of a strong pressure gradient moving across the area. And more than anything else, the weather is the one contingency that would stop the Mount Rushmore fireworks display dead in its tracks. Specific temperature, humidity, inclement weather, and wind criteria had to be met before the fireworks show would proceed.

Weather-wise, most weather conditions fell well within acceptable criteria that early summertime afternoon. The temperature was in the 70s, widely scattered thunderstorms dotted the horizon east of the Black Hills and moving away, all of this with above average humidity. The one fly in the ointment had to do with wind.

At the surface, the gentle winds cooled the crowds. Not so on top of George Washington's head. There, the winds were strong and relentless, blowing around our former president's granite hair at speeds outside the criteria to proceed with the fireworks display.

Fortunately, on that day so many years ago, as forecasted, the pressure gradient relaxed as evening fell in plenty of time to allow the hundreds of thousands of spectators — whether on-site, off-site, on television or the Internet — to enjoy this Fourth of July spectacular show from this most appropriate location, the Shrine of Democracy in the beautiful Black Hills of South Dakota.

'Snow Eater' Winds Push Area Toward Warmth

The jet stream giveth, and it taketh away. Even during those warmer than typical winters, changes in the jet stream can still giveth freezing cold daily temperatures. A slight southerly nudge from this stream of fast-moving air can and does dislodge frigid air masses from the Arctic tundra, pushing them across the Northern Plains to rudely disrupt a once-mild winter. And February appears to be the month when these jet stream nudges become more frequent.

On one of those occasions, February 14, 2003, one of those mounds of dense, cold Arctic air raced out of Canada, spreading its sub-freezing temperatures across South Dakota. The Rapid City Regional Airport temperature fell to a new record low, bottoming out at 17 degrees Fahrenheit below zero. Downtown Rapid City tolerated a low temperature of 11 degrees Fahrenheit below zero, and Custer reported a low temperature of 25 degrees Fahrenheit below zero. Some locations in the Black Hills reported temperatures as low as 33 degrees Fahrenheit below zero.

It never seems to fail that once ice-cold air of record proportions settles in over the Black Hills, thoughts quickly turn toward those milder, gentler temperatures a few days back. No doubt, warmer temperatures will return; but that seems unimaginable as you peel off layer after layer of body-protecting clothing after coming inside from the frigid cold outside.

Imagine the despair when the temperature reached 57 degrees Fahrenheit below zero in McIntosh, SD on February 17, 1936, or, closer to home, 34 degrees Fahrenheit below zero on February 11, 1899, in Rapid City. Inevitably, some residents back then concluded that only the fires of hell could ever restore warmth to the region.

Fortunately, nature does indeed take away the bitter cold's sting so often recorded during February by offering several mechanisms that start the warming process post-haste, if and only if, there was no snow on the ground. The day-to-day increase in the number of sunlight hours promoting notable warming kicks into gear big time during late February and early March. No doubt, more and more minutes of sunshine will eventually eat away that chilly visitor from Canada.

Unfortunately, widespread snow on the frozen ground slows this warming process. Instead of the sun's energy warming the air temperature, the highly reflective characteristics of snow (remember albedo) returns much of that warming energy into space. What little energy/heat remains to impact the snow will be used up working to melt or even evaporate the snow and not warm the air. First things first, get rid of the snow on the ground, so warming of the air can begin in earnest.

Turning the winds across South Dakota around from northerly to southerly, by sliding this large air mass of coldness into Minnesota, should also start the warming process. Regrettably, southerly winds may not always offer up warming winds. You see, as the Arctic air mass spreads southward and eastward, those supposedly warm southerly winds would only return yesterday's cold air back across those snow-covered fields. Again, get rid of the ground-locked snow, then increasing sunlight or southerly winds, or both together, would more quickly jump-start the warming annd melting processes.

Snow covered ground in the Black Hills, SD

Our best "snow eater" to take away the snow covering the ground and take away those bitterly cold temperatures, starts in the Pacific Ocean. Over the Pacific Ocean, westerly winds carry the warm, moist air toward the Pacific Northwest coast. After the moisture-laden air reaches the coastline, it encounters numerous mountains and hills including the Rocky Mountains aligned along Western Montana, British Columbia and Alberta, the Big Horns mountains, and the Black Hills. These mountain ranges force the air upward, causing it to expand and cool, wringing out moisture that falls to the ground as rain or snow.

On the down-sloped leeward side of these mountain ranges, the downward-traveling air parcel contracts and warms, traveling from low to high pressure, releasing back to the surrounding environment the heat/energy once needed to keep the air molecules expanded. The warming relief, known as the "chinook" or "foehn" winds, or just plain "snow eaters," keep our South Dakota winters somewhat manageable.

The now much warmer and drier air sweeps across the state, with its most notable warming effect, felt here in the Black Hills region. These chinook winds, in turn, rapidly eat away the snow on the ground as well as warm the air, setting up periods of rapid warming.

Chinook wintertime effects include the rapid temperature increases, rapid snowmelt, loss of humidity, and fast melting of ice on lakes and rivers. The outcomes from summertime chinook winds frequently occur on the southern and eastern slopes of the Black Hills and downwind from the Black Hills. The drying effects of these winds can result in violent grass fires or forest fires, damage to small plants and cause the loss of soil moisture. We gladly welcome the occasional wintertime chinook visitor, not so much the summertime chinook visitor.

Snowmelt from Chinook Winds a Sublime Process

As discussed previously, chinook winds help to take the big chill out of the Black Hill's winter season. So now, I would like to expand on the warming narrative, because it's so prevalent here in the Black Hills. In the words of Paul Harvey, here's the rest of the story.

In addition to heating the environment through changes in water phases from water vapor to water and from water to ice, pressure differences also warm and cool the environment as air parcels move down the canyons and up the slopes of the Black Hills. Envision a parcel of air running up against the western slope of the Black Hills. As the parcel moves upward, the pressure around it decreases — air pressure decreases with height, allowing the parcel of air to expand.

The first law of thermodynamics dictates that the energy in the air parcel can be used one of two ways. The energy can either be used to expand the air parcel or used to maintain the temperature of the air parcel by keeping it at constant pressure — but not both. Because our climbing air parcel uses the energy/heat from the environment to expand the air molecules as the air parcel moves higher up the mountain slope, the environment around the air parcel cools.

Interestingly, staying within the troposphere layer and given no outside source of heat, the rising air's environmental air temperature cools at a constant rate of about 5.5 degrees Fahrenheit for every 1,000 feet it rises. Meteorologists call this change in temperature with height the dry adiabatic lapse rate.

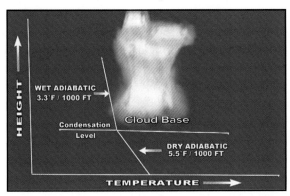

Tropospheric dry and wet adiabatic temperature lapse rate

If the environmental air cools enough, condensation takes place, forming clouds, and precipitation, which in turn adds some latent heat of condensation to the expanding and cooling environmental air. Adding a little bit of heat, due to condensation, to the environment causes the environment to cool with height at a slower rate of 3.3 degrees Fahrenheit per 1,000 feet. Meteorologists call this more gradual change in temperature with height the wet or saturated adiabatic lapse rate.

After the westerly winds pass over the ridgeline of the Black Hills and begin their easterly descent toward Rapid City and the prairies and plains of West River below, their surroundings now exert greater and greater pressure. The added pressure returns energy/heat to the environmental air by squeezing ever closer the disturbed air molecules. Now the environmental air heats up at 5.5 degrees Fahrenheit for every 1,000 feet it drops under the increasing weight of the atmosphere.

As the wind descends from Black Elk Peak (formerly Harney Peak) at 7,742 feet to Rapid City at 3,200 feet or a 4,542 feet difference in elevation, the downward-moving environmental air would warm approximately 25 degrees Fahrenheit. To better illustrate, environmental air with a 20-degree Fahrenheit temperature crossing over Black Elk Peak can warm to 45 degrees Fahrenheit by the time it reaches Rapid City. Not only is it a much warmer air mass when it reaches the eastern foothills and plains, but also a dry air mass. First condensation wrings out much of its moisture climbing the windward side of the Black Hills, followed by a notable reduction of its relative humidity due to the warming descending process.

Now, as this warm and dry air mass slides across the snow-covered ground, it not only warms the outside temperature but also "eats away the snow." The warming effect of chinook winds can rapidly transform the snow from a solid (ice) into a liquid (water), as well as from a solid (ice) directly to a gas (water vapor), bypassing the melting stage. Meteorologists call this process sublimation. You can see the sublimation process in action merely by watching a piece of dry ice vanish into thin air without leaving a wet spot.

So now you have heard the rest of the story.

Dan Lutz

From these bulbous mountains of dark clouds come the most intense downpours, distant cannonades reverberating on the horizon, nonstop gigantic electrical sparks of energy, hail the size of baseballs and our atmosphere's most savage creature: the tornado.

🍂 PRECIPITATION & THUNDERSHOWERS

A rain shower metamorphosis takes place typically during early spring when those general winter storms producing steady precipitation across vast expanses become more localized in nature. As the calendar transitions from winter to spring, the precipitation falling from the horizontal wispy stratified clouds transition into more showery type precipitation associated with vertically developing cumulus clouds that billow up over limited areas.

This transition happens during early spring because the sun's energy rays impact the Earth's surface more directly and for a longer duration as the sun climbs higher in the afternoon sky. The increase in energy equates to a substantial increase in heat per unit of the area when compared to winter's lackluster energy output.

With the upper levels of the atmosphere still cold and the near-ground layers of air progressively heating up, the atmosphere becomes more unstable and begins to boil. The boiling action sends vertical columns of air soaring skyward, cooling and, in turn, condensing to form cumulus clouds. Fair-weather cumulus becomes towering cumulus, and if conditions are favorable, these clouds become rain shower-producing and possibly the more powerful thunderstorm-producing cumulonimbus clouds.

The early-spring time frame is when rain showers begin to fall to Earth, and perhaps thunder and lightning can be heard and seen streaking across the horizon. The largest and most intense thunderstorms occur when a group of cumulonimbus clouds lines up to form into a squall line or gather together to become a multicellular storm or "supercell." From these bulbous mountains of dark clouds come the most intense downpours, distant cannonades reverberating on the horizon, nonstop gigantic electrical sparks of energy, hail the size of baseballs and our atmosphere's most savage creature: the tornado.

No other atmospheric convulsion is more terrifying, and consequently more intriguing than the tornado. The corkscrew spinning appendage from beneath a murky sky surely must stand as the most alarming spectacle nature can concoct. Luckily, most of us will never experience the reverential fear brought about by witnessing the black column of air slinging topsoil and wreckage in all directions. Though tornadoes affect relatively small areas, they strike faster and more swiftly than any other storm.

What is Normal Precipitation?

Rapid City generally gets less moisture than other communities in South Dakota. During wet years, the talk emphasizes above-normal precipitation and conversely, during dry years, the news and political talk stresses below-normal moisture here in South Dakota. It begs the question to define "normal precipitation."

Instead of discussing monthly average precipitation, I will talk only about normal annual precipitation, or the amount of moisture any location might receive each year. First, the term precipitation includes all moisture falling from the clouds, whether liquid or ice. Rain, hail, sleet, and snow all fall (pardon the pun) under the category of precipitation. The term "normal" means the measured precipitation reaching the ground averaged over 30 years.

Normal precipitation figures update every ten years by replacing the starting decade of precipitation data with the most recent decade precipitation data and averaging again over the new 30-year period. For the Rapid City airport, the normal annual precipitation is on the order of 18.32 inches. All normal precipitation records include snow and ice measurements by simply melting them into liquid form. Compared to other large communities in South Dakota, Rapid City finds itself on the short end of the rain/snow stick. On average, Aberdeen receives about 21.76 inches annually, Huron gets about 22.87 inches, and Sioux Falls enjoys approximately 26.35 inches of precipitation each year.

Average Annual Precipitation (1981 - 2010)
(National Climatic Data Center)

East River Communities	Annual Precipitation (Inches)	West River Communities	Annual Precipitation (Inches)
Aberdeen	21.76	Custer	19.59
Brookings	24.24	Deadwood	28.40
Huron	22.87	Hot Springs	17.66
Sioux Falls	26.35	Lead	30.48
Vermillion	27.64	Philip	17.06
Watertown	21.81	Interior	18.46
Yankton	27.09	Rapid City	18.32
Pierre	19.95	Hill City	20.88
Mobridge	16.30	Lemmon	18.10

In the bigger scheme of wetness, South Dakota regularly ends up with less moisture than most areas in the country. The East Coast, the Great Lakes region, the Southeast, and South generally attract more annual precipitation than here in the Great Plains.

Anywhere east and south of Rapid City in the contiguous United States will record higher than normal totals of the wet stuff. For example, Minneapolis receives 28.32 inches, Chicago gets 35.82 inches, Washington, D.C., receives 40.24 Inches, New York collects 41.59 Inches, Oklahoma City gets 33.30 inches and Miami receives 55.91 inches of precipitation.

This wet-weather pattern also suggests that regions closer to large bodies of water, the Atlantic and Pacific Oceans and the Gulf of Mexico, should receive more annual rain or snow than areas located in the interior of the country. In the United States, the Northern Plains finds itself as far as one can get from a large body of water, which suggests one reason for its relatively dry climate.

This closer-the-wetter rationale breaks down slightly for areas between Rapid City and that large body of water known as the Pacific Ocean. Out West, there exists a mixed bag of average annual precipitation totals. Cities closer to the Pacific Ocean than Rapid City, such as Denver CO, Helena, MT, Cheyenne, WY, Salt Lake City, UT, and Los Angeles, CA., all receive less annual precipitation when compared to Rapid City.

Apparently, factors other than proximity to water come into play. The terrain greatly influences the moisture distribution across the West. The Rocky Mountains act as a moisture block or a precipitation/rain shadow across many cities, wringing out much of the Pacific moisture from the eastbound river of air as it climbs and condenses over the mountains.

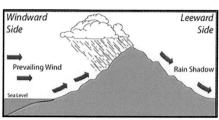

Rain shadow effect (Yahoo Images)

A clearly defined precipitation shadow exists along the leeward side of the Rocky Mountains, enveloping many mountain cities, including Las Vegas, Denver, Salt Lake City and, to some degree, Rapid City. Additional factors, such as prevailing winds, stability and the location of the jet stream, keep other western seaside localities relatively dry. San Diego and Los Angeles both receive less annual moisture than Rapid City, mainly because the jet stream (the highway of storms) blocks many of those rain-making storms from entering their city limits.

If the annual precipitation normal remains consistent from year to year, life adjusts accordingly. It's when the average precipitation becomes the exception that stress and hardship begin to raise their ugly head, no matter where in the country.

Nature Has Own Sprinkler Track

Lawn sprinklers of all shapes and sizes work overtime during hot and dry summers. Some sprinklers wave water drops back and forth in high arcs, and others twirl water drops around and around while standing in place, and still, others stick their spouts out of the ground to wet grass for a period and then quickly disappear.

My personal favorite sprinkler system is the "tractor sprinkler." No need to move sprinklers around the yard every half hour or lay expensive under-ground-irrigation pipes. With the tractor-sprinkler system, you decide on the path you want to be watered and place the hose, which serves as rails such as those on a railroad track, along said path. By straddling the tractor-sprinkler on the hose, it will follow this predetermined path across those areas in desperate need of water.

Nature also lays out a track for its sprinkler system, but unlike our backyard sprinkler, we cannot steer nature to lay the track over those areas in most need of water. Nature locates its storm track preferably along the outer edges of high-pressure systems. And because of the pre-determined airflow around the high-pressure system, these tracks tend to follow a clockwise direction. One of the main summer tracks for nature's storms follows the outer edges of the Bermuda High-pressure area.

The center of the semi-permanent Bermuda High-pressure area dwells in the Atlantic Ocean east of the Carolinas near the Island of Bermuda. This considerable weather system sits year-round over the Atlantic waters at about the 30 degrees north latitude. It elongates westward across the Eastern U.S. during the early summer months, only to withdraw back to the Atlantic in the fall. This semi-permanent high-pressure system mainly influences the weather east of the Rocky Mountains.

Location of the semi-permanent Bermuda High-pressure system

Depending on its strength and size, the clockwise track around this dome of air directs copious amounts of rain from the South Atlantic eastward across the Gulf of Mexico and into the Southern and Central Plains. These bands of wet storms that march across the Gulf of Mexico, can easily cause rivers and streams in Central Texas to rise rapidly 20 to 30 feet above flood stages.

The clockwise track surrounding the Bermuda High becomes less northerly and more easterly across the Northern Plains, typically directing storms from Colorado and Wyoming eastward across the Dakotas and the Black Hills.

Areas of the country east of the pathway of storms and inside the sprinkler track of wetness, often experience many weeks of dangerously hot and humid weather, referred to as the "dog days." The Midwest to the Mid-Atlantic States suffer the most dog-day afternoons borne from the Bermuda High-pressure system, but seldom are they a concern in the Black Hills region.

On occasion, when unseasonably hot and dry periods do occur here in the Black Hills, we too will feel the impact under an elongated piece of the Bermuda High reaching well into the Northern Plains region pushing the storm-following track farther west into Wyoming and Montana and north across Southern Canada and away from the Black Hills. Once the Bermuda High-pressure area relaxes its grip on the Northern Plains, the usual storm tracks return across South Dakota.

Rain and Mud Follow in Fire's Path

Fire danger increases notably during those dry and hot days that on occasion best characterize the summers here in the Black Hills. During the driest of summers, a week does not pass without above-normal temperatures or rainless skies, and often, skies filled with smoke and soot carrying away some of the beauty of the Black Hills. In summers past, charred wood, soot, and ash from forest and grass fires have fallen on Rapid City like a summer snowfall.

Past wildfires, such as the 2017 Legion Lake Fire that torched 35,000 acres in and around Custer State Park and the 2000 Jasper Fire that scorched 83,510 acres across the Southern Hills. Other notable wildfires include the 2012 Oil Creek Fire that burned 63,140 acres of Wyoming's Black Hill forest, the Grizzly Gulch Fire, the Little Elk Fire, and the Battle Creek Fire to name only a few of the more recent fires. These fires spread fast-moving flames across dried-up grassland and prairies, and high into trees in search of more Black Hills beauty to devour. Large forest fires can easily reach temperatures well into the 1,000-degree Fahrenheit range, wiping out all plant life in its path, cremating seeds buried deep in the ground and hardening the soil.

Flames reaching skyward from the 2000 Jasper fire, Custer County, SD (U.S. Forest Service)

After the fires have died and the ground has cooled, the devastation of the now-blemished Black Hills lays barren along many of the main gateways into the Black Hills. Keep in mind that even though the more intense fires bring total ruin, the many lesser intense fires begin the cycle that reinvigorates the soil and new plant life to start the renewal of the Black Hills.

Years later after the fires, a ride along I-90, South Dakota Highways 16, 44 and 79 or through Boulder Canyon toward Deadwood, the aftermath of these fires, still can be seen. The blackened, hardened and defoliated steep canyon walls that once provided shelter and character to so many Black Hills communities have become a breeding ground for yet more ruin. The now impervious soil becomes the breeding ground for severe erosion, often polluting our streams and creeks with silt and other oxygen-choking material.

House damage from the 2002 Deadwood, SD mudslide (U.S. Geological Service)

With no lush grass, thick bushes or brimming trees to slow the progression of runoff, even the typical short-lived afternoon rain can readily cause flash flooding and mudslides. Teresa Murphy, a former hydrologist with the Rapid City National Weather Service, says flash floods can happen over burned hillsides with as little as one-tenth of an inch of rain per hour. Mudslides can occur with a half-inch per hour of rainfall.

On August 10, 2002, an early evening line of thunderstorms moving from west to east at 35-40 miles per hour dropped about four-tenths of an inch of rain in 15 minutes on top of a burned-out gulch overlooking Deadwood's hospital. Within 12 minutes, mud and debris toppled from the hillside into parts of Deadwood spreading sludge around the hospital, and covering streets and sidewalks, damaging houses, cars, and motorcycles.

When forest fires happen in hilly, steep terrain, it's not a question of whether mudslides will occur — it's all about when they will happen. Unfortunately, many communities located at the base of steep terrain in the Black Hills spared from the devastation of wildfires confined to forests above, can almost certainly count on sludge and rocks to move down from burnt hillsides with the next rain.

Jasper Fires Create Rain

Shape, size, and location of clouds represent the outgrowth of atmospheric convolutions. With every mood swing of the atmosphere, the clouds respond almost immediately. When high cirrus clouds blow in and thicken from the northwest, it suggests that the afternoon's dry tranquil atmosphere may change into a wet turbid sky, foretelling a fast-moving line of thunderstorms rolling across the Black Hills. Fair-weather late afternoon puffy cumulus clouds assure us a stable atmosphere and that the day will remain pleasant and absent of any torrential downpours. Day-long "sheet-like" stratus clouds often foretell a calm atmosphere and a quiet evening.

Vertical currents of air, of various speeds, start the cloud-making process. Cold air over hot air causes instability of the air column, moving the air upward to initiate the beginnings of cloud-making. As the air rises, moving to a lower pressure environment, it cools condensing the water vapor into trillions of tiny droplets in the rising column. Those massive amounts of minute water droplets trapped in the rising pillar of air make up the fabric of those sometimes white, other times dark clouds, sliding across the sky.

It's those atmospheric mutations well above the ground that form clouds. On occasion, however, Earth-bound catastrophes can trigger overhead cloud development, even under stable atmospheric conditions. Volcano eruptions can create large clouds by abruptly pushing the air located over the eruption site skyward, causing the air to condense on copious amounts of soot and ash spewing out the volcano crater. As the cloud drifts downwind, these tiny cloud drops can collide with one another forming muddy raindrops that fall to the Earth hundreds of miles downwind.

Large forest fires can quickly produce fast-growing clouds (pyrocumulus) and subsequent rainfall within a stable atmosphere. The 2000 catastrophic Jasper fire offered up its own fast-growing cloud, and in turn, a heavy rain shower with accompanying strong wind gusts felt miles away.

Formation of a Jasper fire rain cloud, Black Hills, SD (U.S. Forest Service)

This relatively small area of intense fire generated substantial instability coupled with rising smoke and soot particles from burning trees and brush forming this now large dark-gray rain cloud. Folks attending the Dakota Thunder 2000 Air Show and Open House at Ellsworth AFB that day, wondered why such an ominous looking cloud developed over the Southwestern Black Hills on an otherwise sunny afternoon.

The cloud remained stationary over the fire area as the smoke particles continuously fed into its updraft allowing for substantial growth. Finally, after a period of a few hours, the intense falling rain triggered a strong gust front to propagate from the Southern Hills to and across Ellsworth AFB, sweeping away many empty cups, boxes, and displays.

The effects of the Jasper fire were not merely confined to the blackened footprints scarring our beautiful Southwestern Black Hills. The upper-level and surface winds generated by the fire carried the smoke and soot hundreds of miles downwind across State Dakota and beyond.

Some Names Pack Punch.

What's in the name of a hurricane depends on the context. Specific names represent the destruction of homes, businesses, and lives. Andrew, Camille, Hugo, Katrina, and Harvey are names that represent such devastation and hardship, particularly along the Atlantic and Gulf of Mexico coastlines. These names, of course, represent only a few of the most notorious hurricanes that, in the past, severely impacted this country.

As far back as records can take us, hurricanes have unleashed their wrath along mostly the eastern and southern shores of this nation. And, these same records indicate that the act of assigning names to these most powerful of all storms dates back several hundred years.

Satellite photo of a massive hurricane with eye clearly visible (National Hurrican Relief Center)

Saints, girlfriends, years, the phonetic alphabet (Able, Baker, Charlie) and even names of first ladies found their way in identifying hurricanes. Case in point, "Hurricane Bess" was named after President Harry Truman's wife. In the West Indies, on September 13, 1876, a hurricane struck Puerto Rico on St. San Felipe Day. Not too surprisingly, the people named that hurricane "Hurricane San Felipe." Interestingly, nearly 50 years later, on San Felipe Day, another hurricane struck Puerto Rico and was given the name "Hurricane San Felipe the Second."

In the past, the use of latitude and longitude readings to identify storms often caused confusion and even heightened false rumors when critical information broadcast from one radio station about one storm was mistaken for an entirely different storm hundreds of miles away. In 1953, U.S. hurricane officials realized that giving storms short, distinctive names in both written, as well as spoken communication, lessened confusion, greatly benefiting widely scattered weather stations, coastal bases, and ships at sea.

Also, in 1953, the naming convention restricted the hurricane designate to female names only. This practice of solely using female names came to an end in 1979, when the National Hurricane Center (NHC) used both female and male names for all Atlantic and Gulf of Mexico tropical storms and hurricanes. When a tropical disturbance over the Atlantic or Gulf of Mexico intensifies into a tropical storm — with rotary circulation and wind speeds above 39 miles per hour — the NHC will give the storm a name from that year's alphabetic list of predetermined names beginning with the first name on the list. The boy or girl names would alternate down the register and from year to year.

The list of names repeats itself every six years. The year 2020 list of names will be used again in the year 2026. So, it is conceivable that the name Lili, used to identify a 2002 hurricane, could have been used to identify yet another hurricane in 2008. Some names never will make a list, mostly those names starting with the letters Q, U, X, Y, and Z, due to the scarcity of names beginning with those letters. Also, names of hurricanes having a significant economic impact or loss of life will have their names retired.

From 1953 through 2018, 86 Atlantic hurricanes' names now exist only in retirement, including Katrina and Harvey, two of the most expensive ($125 billion each) in U.S. history. No future hurricane will have the name Camille, the third strongest hurricane to hit the U.S. in the 20th century, or Hugo, one of the strongest storms to ever hit the East Coast. Recently the World Meteorological Organization voted to retire the names of two 2018 hurricanes — Florence and Michael. By the way, big bad Bob retired in 1991.

Nature's Own Light Show

What can be more spectacular than a springtime thunderstorm igniting over the Black Hills? A thunderstorm, by its very name, tells you it has thunder, and where there's thunder, there's always lightning nearby.

In-cloud and cloud-to-ground lightning, Rapid City, SD (NewsCenter1)

Lightning strikes somewhere on the Earth's surface about 100 times every second, or about eight million strikes per day. Interestingly, most lightning does not reach the ground. Some bolts of lightning never leave the confines of the thunderstorm itself, while others leap from one cloud to another cloud. Lightning is the big brother of that spark of electricity you sometimes feel when you scuff your feet on a rug on a low-humidity day and touch a doorknob.

Like that static spark, lightning lasts only a fraction of a second. Unlike that annoying spark, lightning generates an immense amount of power and heat. Both static sparks and thunderstorm lightning occur because of the attraction between opposite charges.

Strong air motions within the thunderstorm cloud carry, split and collide water drops and ice crystals among themselves, forming a skewed distribution of electrical charges. Typically, a high concentration of negative charges ends up near the base of the thunderstorm, and the opposite positive charges find themselves near the cloud's top. As discussed previously, nature abhors discord and, therefore, will periodically discharges lightning to align better these wayward electrical charges.

Those of us Earthbound inhabitants need not fear the in-cloud or cloud-to-cloud lightning variety located safely far above our heads, but we should fear the type that strikes the ground or cloud-to-ground lightning. Deaths caused by tornadoes and floods far exceed annual lightning fatalities in the United States, but lightning deaths do happen and those numbers remain close to hurricane deaths.

So, how does that bolt of lightning reach the ground anyway? Besides the concentration of negative electrical charges in the bottom of the cloud, a corresponding grouping of positive charges builds upon the skin of the Earth.

Negatively charged electrons begin spilling out downward, attracted by the positive charges on the ground. This downward movement is called the "stepped leader." As the step leader descends, the ground-based positive charges begin to ascend from the Earth. This ascension of positive charges is called the "streamer." When the stepped leader and the streamer meet, they form a path or temporary channel through the atmosphere that conducts electricity efficiently.

We see the channel conducting that electricity as a bright and scorching display of light flickering as the primary and return strokes flash back and forth. A typical lightning stroke may last only 30 milliseconds. Four or five lightning strokes within the path can and do occur in a blink of the eye.

And regardless of the old saying, lightning does strike the same place twice or even more. So, this spring, when nature reaches down to touch Earth's doorknob, enjoy its beauty and majesty cautiously, from a safe distance.

The Genesis of Thunderstorms

The casual observer cannot fully comprehend the violent turmoil within the white foamy exterior of a massive thunderstorm. Lightning is but one of the myriads of natural by-products found within these behemoth clouds that begin their life cycle as one of those cotton-like fair-weather cumulus clouds dotting the afternoon spring sky, only to grow further becoming a towering cumulus and then on to

Fair weather cumulus clouds floating over Rapid City, SD

becoming a full-blown thunderstorm. These dynamos in the sky also conceive powerful tornadoes, crop-destroying hailstones, dangerous flash floods, and destructive straight-line winds.

The cloud itself and its violent by-products exist because of upward-moving air currents within the cloud. The initial updraft covers an area typically less than a mile in diameter, as the air begins its ascent at a vertical speed on the order of 15 to 30 miles per hour. Different heating characteristics of the Earth's surface, due to differences in composition, color, shape, and texture, all feed into the making of these early-on vertical moving columns of air.

For example, a field of freshly plowed soil can capture more of the sun's heat than a nearby pasture of green grass. The warmer plowed field, in turn, can more efficiently heat the air overhead, causing it to expand and become less dense sooner than the relatively cooler surrounding air. Less dense air means more space between molecules, making the air relatively lighter and more buoyant.

A towering cumulus cloud growing over the Black Hills, SD

The heated and lighter air begins its rise away from the Earth's surface, while its neighboring, cooler and denser air and, consequently, heavier air sinks and rushes in to fill the void near the surface left by the updraft. Vertical currents of air begin to cavort across the warm spring afternoon skies, some moving upward, and others rushing downward toward Earth. Soaring birds of prey gaining altitude overhead for a better view of their hunting ground below signal the presence of these vast columns of rising air.

The upward flowing air column frequently contains substantial amounts of water vapor, invisible to the eye until the colorless gas condenses into its liquid state, forming cloud droplets hundreds or thousands of feet above the Earth's surface. Condensation occurs when the rising air cools to the point or altitude where the humidity reaches 100 percent, and the rising air column can no longer maintain its moisture in its gaseous state. Further rising and subsequent cooling allow nature's thermodynamics to "squeeze out" the vaporous moisture to form the liquid cloud droplets. The always present, ever-so-tiny bits of solid particles in the air, such as dust or salt, serve as collection points for the small cloud droplets. At frigid temperatures, the cloud droplets can quickly freeze into ice crystals.

Typically, the condensation process takes place at the same altitude throughout the rising column causing clouds to have a relatively flat and uniform base. Nearby clouds follow the same recipe, using primarily the same ingredients, resulting in a standard cloud base height among neighboring clouds.

Given the right atmospheric conditions, a weak vertical puff of air can rapidly and steadily expand in both size and strength. The birth of these rapidly rising clouds once started, does not assure maturity. Totally dependent on the expanding cloud's environment and available water vapor, most late afternoon clouds never become thunderstorms due to unfavorable atmospheric conditions.

Thunderstorm Adolescence

Change, challenge, and new adventures in living best describe the teenage years. Like people, the growing cloud's adolescence stage of development best defines the thunderstorm's time for a transformation of monumental proportion.

Those of us who live around teenagers know that they hardly exist in the morning, but spring to life by late afternoon, with activity reaching a peak during the evening hours. Their rate of growth no longer appears gradual but in quantum jumps of height and weight. Their internal emotional swings continually churn, one moment reaching new heights, and the next dropping to new lows. And, like the teenager's metamorphosis, soft and docile white puffy clouds that pop-up across the late morning sky can quickly manifest themselves as torrents of afternoon weather mutations.

This sudden change in demeanor results from an explosion of acquired energy bursting from within the water vapor trapped in the cloud's updraft. Adolescent individuals get their energy outbreaks from junk food, quickly washed down with a soft drink. Adolescent clouds obtain their energy bursts from the water vapor captured and carried to the cloud by persistent updrafts. The process by which the water vapor releases its enormous amounts of heat and energy in the cloud's unsuspecting adolescence is our old friend "latent heat of condensation."

As the cloud's updraft carries these agitated water vapor molecules up and into the cloud base, they cool and slow down, transitioning from water vapor, a gas, to cloud droplets, a liquid. The water vapor molecules do not change, they remain H_2O, but their speed of movement slows down. So, after condensation, these water molecules find themselves closer together, moving slower,

Passing rain shower over the South Dakota grassland (South Dakota Department of Tourism)

releasing the energy/heat agitating the molecules necessary to keep them apart, thus warming the environment. The energy/heat escapes into the environment, adding more buoyancy to the developing cloud, rapidly increasing its height and strength, pushing the adolescent cloud toward maturity.

Within the growing adolescent cloud, the additional liquid water droplets formed by the ongoing condensation process attach onto older cloud droplets, causing those droplets to grow. Because different cloud droplets grow at different rates, collisions among cloud droplets abound within our teenage cloud. The slower-moving larger droplets become even larger due to the collisions from the faster-moving smaller droplets.

Eventually, the larger droplets become large and heavy enough to overcome the lift from the cloud's updraft, and rain begins its journey to the Earth's surface.

The taller these clouds grow, the faster the updrafts become, processing more water vapor, more heat and, subsequently, allowing the cloud to become more buoyant, pushing cloud tops to heights up to 20,000 to 25,000 feet above sea level.

When the cloud growth extends higher than the freezing level, the cloud droplets freeze to become small ice pellets, and some of the moisture changes from water vapor directly to ice crystals (desublimation). Both the liquid-to-ice and water vapor-to-ice freezing transformation processes accordingly releases even more latent heat into the growing and strengthening updraft. The sharp, defined edges of the teenage cloud indicate the portion of the cloud that remains in a liquid state. The fuzzy and less-defined edges often found near the top of the cloud suggest the presence of ice crystals.

Our fair-weather puffy cumulus cloud is now a rain shower. A rain shower, however, does not make a full-blown thunderstorm.

Ups, Downs of Thunderstorm Maturity

The term "anthropomorphize" means to apply human characteristics to inanimate objects. I did precisely that in comparing the early life of a spring thunderstorm to adolescence. Only after that intermediate step in development can we begin to appreciate the enormity that awaits when the thunderstorm grows to full size.

A mature thunderstorm with an anvil top (Google Images)

When an adolescent thunderstorm matures, its presence reaches from just a few thousand feet off the ground up to the very top of the troposphere. It is at these heights, 50,000 feet to 60,000 feet (10 miles or more) when the now fully developed thunderstorm spreads out its characteristic white fuzzy-pointed finger that's so common during a late spring and summer thunderstorm.

This broad expanse of frozen crystals, often spreading southeastward here in the Northern Plains, warned many a farmer or rancher of the impending onslaught of heavy rains, strong winds and large hail in the days before computer model forecasts.

Thunderstorm maturity is that time of life when both updrafts and downdrafts reach their ultimate strength. Downdrafts gain intensity, in part, from frictional dragging induced by falling raindrops and hail. Large raindrops just outside the updraft fall headlong toward Earth at terminal velocity, dragging with them the surrounding air molecules. As the rain dives just outside the updraft, it begins to evaporate. Evaporation, when the water liquid becomes vapor, uses up energy/heat from the surrounding air, cooling the air and making it denser (just like your body cools off when getting out of a shower or pool). This increase in density or weight causes the downdraft rush to plummet even faster toward the fields and homes below.

The sinking air speeds toward the Earth until it smashes into the ground, but the energy force does not stop there. It's redirected in the only direction it can go, outward, with a sudden rush of cold air spreading out in all directions. The resulting rush of outward traveling air, better known as a gust front or micro-burst, quickly moves away from the mature thunderstorm cloud that caused it. This early strike of strong chilly gusts that originated from the thunderstorm miles elsewhere disrupts the calm before the storm. These quick micro-bursts can be worse than the thunderstorms from which they came. Gust fronts emanating from the most severe thunderstorm can and often do exceed hurricane-force wind speeds.

Gust fronts that precede the parent thunderstorm perform like a small-scale cold front, themselves triggering a line of thunderstorms in advance of the parent thunderstorm. The boundary of this violent storm breakout, the squall line, often produces the most severe thunderstorm activity, i.e., damaging hail and tornadoes.

And these outflow boundaries do not necessarily vanish with distance traveled or the setting sun. Often, they continue to propagate undetected overnight hundreds of miles away from their source, ready to trigger secondary thunderstorms the following day. Like grandchildren and great-grandchildren, thunderstorms can return the next afternoon along the downburst outflow boundary as subsequent generations of the original mature thunderstorm.

Straight-Line Winds Blamed for Damage

During the evening of August 1, 2000, an intense downburst with wind gusts higher than 100 miles per hour blew through Spearfish, SD, severely damaging homes, businesses and the landscape throughout this Northern Black Hills community. And a few days later, a similar weather event blew through Mitchell, SD, where ten people were injured.

Some said a tornado, and others said straight-line winds from powerful thunderstorms. The official report from the National Weather Service (NWS) blamed straight-line winds, caused by a series of downbursts from a line of powerful thunderstorms, for the extensive damage.

Here's the low-down on this unusual weather phenomena so you can judge for yourself. First, a line of thunderstorms in question raced toward Spearfish from the northwest at speeds of up to 60 miles per hour. This speed alone gives any stationary object in the path of these fast-moving thunderstorms a 60 mile per hour smackdown. It's like trying to catch a ball from someone in a car moving down the road at 60 miles per hour. The toss from the speeding vehicle may be gentle enough, but that ball will hit you at roughly 60 miles per hour.

Secondly, the might of the thunderstorm comes from powerful updrafts and downdrafts. As discussed previously, massive amounts of falling rain and hail literally can drag the surrounding air to the ground at tremendous speeds. The harder the rain falls, the stronger the downdraft, and the stronger that downdraft hits the ground, the higher the wind speed radiates out from its epicenter. These winds are known as "straight-line winds."

Thirdly, you have heard talk about "dry thunderstorms," regarding the Western State's fires. A dry thunderstorm means the excessively dry air beneath the thunderstorm evaporates the rain before it reaches the ground. The lightning, however, continues its Earthbound journey triggering many wildfires. That evening, a relatively thin surface layer of dry air spread across Spearfish, causing much of the rain falling from those fast-moving thunderstorms to evaporate before reaching the ground. The evaporation drastically cooled the surrounding air, making it even heavier than its surroundings, and causing it to plummet to the ground with even faster speed. Finally, the fast-moving downdraft smashes into the ground and spreads out and away from the storm carrying destructive wind speeds.

A preliminary study of the damage by the Rapid City NWS showed a uniformly spewed debris field spread out toward the southeast. A consistent pattern of loss in one direction often suggests straight-line winds. Tornado damage, on the other hand, would have been spread over a broader range of tracks, with structures/trees generally blown toward one another.

On that evening, a fast-moving line of thunderstorms, coupled with powerful downdrafts, created straight-line wind gusts of more than 100 miles per hour. Other weather forces besides tornadoes can and do yield tornado-like wind speeds, such as those powerful winds that blew through Spearfish and Mitchell, SD.

Giving "Chicken Little" some standing, the sky does fall — sometimes.

The Birth of a Tornado

Tornadoes occur all over the world. These destructive forces of nature occur most frequently in the United States and most often east of the Rocky Mountains during the spring and summer months. The most favorable real estate for tornado development stretches northeastward from north Texas across Oklahoma, Kansas, into Southern Nebraska, and eastward into portions of Iowa and Missouri. This vast expanse of the Plains and Midwest represents an ideal tornado breeding ground by its location, between the warm waters of the Gulf of

Mexico to the southeast and the Rocky Mountains to the west. Three distinct airflows converge into this region, known as Tornado Alley, to establish all the atmospheric ingredients necessary for tornado conception.

Tornado (Dan Lutz)

A tornado is a violently rotating column of air pendant reaching the ground from a towering thundercloud. Its evolution in the thunderstorm remains one of several atmospheric phenomena not fully understood. Theory suggests that nature sows the seeds of a tornado well before thunderstorm development. Picture a rapid change in surface wind direction, now level to the ground, due to the proximity of a low- and high-pressure system with an increase in wind speed with height. This situation creates a natural and invisible horizontal spinning effect in the lower atmosphere. Now, introduce into the mix the beginnings of an intense thunderstorm and its accompanying powerful updraft. The growing thunderstorm updraft draws upright the near-ground parallel rotating column of air, which now becomes a more upright and rotating column of air.

With time an area of surface-based rotation that began many miles wide, now finds itself confined within the smaller space, making up the base of the thunderstorm. The now tightly confined air rotation follows the laws of physics and spins faster and faster, just as ice skaters increase their rate of spin by pulling in their arms closer to their body. The once large-scale horizontal spin quickly increases its speed considerably when lifted into the much smaller-scale base of the thunderstorm. Most powerful and violent tornadoes form within this area of trapped rotation protruding from the cloud base, often called "a wall cloud."

A developing severe thunderstorm with a wall cloud over the Black Hills, SD (Dan Lutz)

There are about 800 tornadoes reported nationwide during an average year. Statistics tell us that annually, approximately 80 lives will be lost, with more than 1,500 injuries due to tornadoes. The most brutal tornadoes leave behind tremendous destruction with wind speeds from 205 to 300 miles per hour or more. Fortunately, these examples represent only about 2 percent of all tornadoes sighted. Fortuitously, about two-thirds of all tornadoes develop relatively weak wind speeds, less than 100 miles per hour. The remaining one-third generate more destructive and formidable wind speeds between 110 and 205 miles per hour.

Booming thunderheads with damaging winds and tornadoes occur almost any time of the year, peaking in the late spring when the sun climbs higher and higher into the sky, allowing surface temperatures to rise accordingly. Regardless of the time of year, all the ingredients needed for severe weather can find their way onto the plains of the Midwest, producing many violent rotating pendants of air from the sky.

It's not uncommon for cold and dense air to slide southward out of Canada across the Northern Plains into the Midwest, then plowing into the warm and humid air moving up from the Gulf of Mexico to the southeast. The cold, denser air slides under the warmer air packed full of moisture, lifting and accelerating it upward high in the sky where the transition from order to chaos happens.

The essential ingredients for tornado development

Added to this mix are not one, but two racing upper-level jet streams; the polar jet stream, and the subtropical jet stream, converging overhead. With these four ingredients in place, nature — in its pursuit to create atmospheric chaos — violently releases its raw energy stored within the chemical bonds of water vapor.

The strong winds aloft usher in a layer of dry air cooling the mid-levels by evaporation, enhancing the instability of the atmosphere, allowing the storms to explode upward and outward. Also, the strong winds aloft create thunderstorm tilt, allowing the storm's own downward-moving rain shaft to fall away from its updraft, causing the storm cells to strengthen and grow into supercells, enhancing the likelihood of funnel cloud development. Without the shear or tilting, the thunderstorm most likely will destroy itself early in its development by allowing its own downward traveling rain shaft to fall straight down into the updraft, weakening the updraft and, in turn, the storm.

Thanks to the now more readily available real-time time data and supercomputers to easily and quickly digest the data, forecasting tornado development has improved greatly over that past few decades. Twenty years ago, the lead time for tornado warnings was at most one minute, and ten years ago it was about 5 minutes. Today, the average lead time before tornadoes strike is about 11 to 12 minutes.

It is comforting to know that tornadoes adversely impact people's lives infrequently here in Western South Dakota due to its rural landscape and a lack of high humidity needed to rev up the strong updrafts essential for tornado growth. On average, Western South Dakota can expect about four to five tornados per year, while remaining ever vigilant for the exception. Unfortunately, it takes only one tornado to affect the lives of hundreds of people cruelly. The 1998 Spencer, SD, tornado will remind us of this simple affirmation for many years to come.

The '98 Tempest Over Spencer

I've explained the birth and maturing stages of a thunderstorm: from the fair-weather cumulus cloud drifting quietly across a late summer sky, through the unpredictability of a towering adolescent cumulus cloud, with its blossoming energy, until finally reaching full-growth in the stage known, in the world of clouds, as a "cumulonimbus," or, "Cb." Cumulonimbus clouds, and specifically the "cumulonimbus mammatus," are enormous thunderstorm clouds easily recognized by the cluster of pouches hanging from its underside. And, from these full-blown cumulonimbus clouds, tornadoes form.

A cumulonimbus mammatus cloud over the Black Hills, SD

For a good reason, no other atmospheric cloud mutation terrifies the people of the Plains states more than the tornado. On May 30, 1998, the late afternoon sky over Spencer, SD, deteriorated from what was a deep blue into a dark, ominous gray. A line of thunderstorms approached from the west, drawing-up the thick humidity that hung in the air, converting much of it into enormous amounts of untamed energy.

To long-time residents of Spencer, hardened from enduring decades of tempestuous South Dakota weather, this sudden change in the atmosphere represented nothing more ominous than another in a series of spring squalls, the windstorms that bring about the necessary and always welcomed spring rains.

Expecting an evening of short-lived heavy rains, many of Spencer's households settled down to their usual Saturday evening activities, giving no more than passing concern to the darkening skies. Regrettably, the very enormity of this line of thunderstorms did not allow even the most seasoned of Spencer's residents to appreciate fully. That is, until only a few minutes before a massive funnel moved through town, ever-expanding as it hurled large chunks of what used to be houses and trees into the surrounding countryside.

As the tornadic line moved across the Hanson-McCook County line, a half mile west-northwest of Spencer, residents saw a large black whirling cloud hanging beneath the darkening, inky sky. The now tornadic cloud descended directly toward the Spencer community, home to 320 people. And, how would they know of its approach in time to take shelter? Apparently, and unfortunately, many of Spencer's residents did not know that within a few minutes tornadic winds would obliterate much of their town.

On the way to Spencer, one of five reported tornadoes in the system ripped apart the main power lines leading into town. The power outage not only blacked-out the community's television sets, but it also shut off power to the Spencer Volunteer Fire Department's metal building, inside of which was the button to activate the community's lone warning siren.

The May 30, 1998 tornado approaching Spencer, SD (Google Images)

Many survivors said their first indication of the horror that approached came from merely peering through windows and doorways. That belated realization left only minutes to flee to a shelter, basement, interior room, or another place of relative safety. Some told of having just enough time to get behind a substantial piece of furniture, or to lie flat on the floor.

In less than seven minutes, winds estimated at nearly 250 miles per hour carved a half-mile-wide path directly across Spencer. Six people died and more than a third of the town's population was injured. Virtually every one of the 190 structures in town, including houses, businesses and public facilities, sustained either total ruin or extensive damage.

Unfortunately, even with all our technology, we cannot prevent tornadoes. For us to continue to share the troposphere with massive twisters, we must know where and when they develop long before they land on our doorstep. The Spencer tornado prompted a needed expansion of the National Weather Service Weather Alert Radio service across South Dakota. Today virtually every populated area of the state has access to those radio broadcasts to hear and act accordingly to the latest severe weather warnings.

You can easily purchase a fairly inexpensive weather radio with either battery power, or AC power with a battery back-up in the event of a power outage, as in Spencer. The technology works. The difficulty continues today because approximately 97 percent of the public do not have one of the inexpensive lifesavers. And, even if it will never provide you and your family an actual life-saving severe weather warning, the small dollar investment in a weather alert radio should be worth the price in the peace of mind, alone.

Man's Look Inside a 1951 Tornado Recounted

Tornadoes occur every month of the year and in every state of the nation, with the majority touching down during the period mid-spring to late spring. Most tornadoes develop between noon and sunset, the time of maximum heating from sunlight, and travel from the southwest toward the northeast. However, in South Dakota, many of these powerful twisters move from the northwest toward the southeast. A sustained movement of a tornado from the east quadrant toward the west rarely happens. The fewest number of tornadoes happen between 3 AM and 5 AM.

Individual tornadoes can and do zigzag, move in tight circles, and even become stationary. In general, tornadoes travel along the ground with great haste, with forwarding speeds ranging between 15 and 35 miles per hour. Some of these massive appendages from the sky can move from one location to the next with great haste at speeds as high as 75 miles per hour. While on the other end of the ground-speed ledger, one South Dakota twister remained fixed in a field for about 45 minutes. Consequently, damage from tornadoes comes quickly. Tornadoes destroy by the blasting effect brought on by a sudden increase in wind speed, the explosion due to a sudden drop in air pressure, aerodynamic pressure differences that can lift rooftops and other structures from their foundations, and the result of violent collisions by large flying objects being tossed around by massive wind speeds.

Wind speeds generated by the most dangerous tornadoes can easily lift houses from their foundations and cause them to disintegrate in seconds. Tornadoes carrying wind speeds of more than 300 miles per hour can readily project automobile-size missiles through the air. These potent twisters with this kind of savagery are usually reserved for, "tornado alley," that stretches from North-Central Texas across Central Oklahoma, Kansas and into Nebraska.

Sudden pressure changes have devastating effects on even the sturdiest of homes. Pressure drops to levels never registered by any other atmospheric disturbance can be found within a tornado vortex. These nearly instantaneous drops in the pressure surrounding a house push exterior walls outward, laying bare the house's interior to the ravages of the tornado's winds. Houses disintegrate within seconds after being touched by the most powerful of nature's land-based tempests.

Tree damage by the 2018 Spearfish Canyon tornado

As incredible as this may sound, one person claims to have looked inside the vortex of a tornado and lived to tell about it. In the June 1951, issue of Weatherwise magazine, Captain Roy S. Hall described the inside of a massive slow-moving tornado as it dropped from the sky, disintegrating his home around him. He recalled, "The interior of the funnel was hollow; the rim itself did not appear to be over ten feet in thickness, and, owing possibly to light within the funnel, appeared perfectly opaque. Down at the bottom, judging from the circle in front of me, the funnel was about 150 yards across. Higher up it was larger and seemed to partly fill with a bright cloud, which shimmered like a fluorescent light. It looked as if the whole column was composed of rings or layers, and when a higher ring moved on toward the southeast, the ring immediately below slipped over to get back under it. This rippling motion continued down toward the lower tip. And if there was a vacuum inside the funnel, as is commonly believed, I was not aware of it."

Far more common than an eyewitness account of the inside of a tornado are the eyewitness accounts of the massive destruction of property and lives left behind in the path of nature's most savage attack from the sky.

No Time for 'Chicken Little'

Tornadoes can and do kill and destroy with unmatched efficiency. I cannot overemphasize the need for any public information released about these dark appendages from the sky to be timely and dead-on accurate. The public requires such knowledge to make informed choices about saving their lives and the lives of their families.

Dust Devil (Google Images)

Many stormy weather phenomena look like tornadoes, especially to the untrained eye. Difficulties in discerning a tornado from other low-hanging cloud forms often arise during the adrenaline-inducing first sighting moment. A tornado is not a "dust devil," which is a spinning column of air hardly ever associated with an appendage hanging from a cloud. Neither is a tornado a whirling column of air that stays aloft, known as a funnel cloud. A tornado is not a wall cloud, which can rotate like a tornado, but at much slower speeds. What we do know is that funnel clouds, and wall clouds strongly suggest the existence of a potential tornadic cloud system.

The National Weather Service (NWS) relies on the word of its trained observers, rather than reports from the public, before confirming a tornado sighting. That's because the public's imagination often runs away with itself when sighting what they believe to be a tornado.

I had a call once reporting a tornado that turned out to be an airplane's contrail settling over the western sky. If the NWS (or television stations, for that matter) acted on every reported sighting of a tornado, they would suffer the results of the "Chicken Little complex," named for the fictional character who panicked his friends by broadcasting the impulsive warning, "The sky is falling."

There must be only one voice issuing weather warnings, especially tornado warnings, to avoid confusion. That voice should always come from the office of the NWS first. Its mission is clear-cut even when the weather is not. NWS folks are all trained professionals who have the very latest sophisticated equipment at their ready to keep watch for severe weather 24 hours a day, every day. They are good at what they do. The NWS coordinates and dutifully records all reports of tornado sightings, fact or fiction. The NWS will always have the final word regarding tornado sightings.

Hail, South Dakota

As stated earlier, a severe weather watch means that conditions are favorable for severe weather to develop, and we need to stay alert for fast-changing weather conditions. A severe weather warning tells us that severe weather is imminent or has been confirmed, either by trained spotters or by radar and that we need to take immediate action if we are in the path of the severe storm.

In the spring and summer, most severe weather warnings stem from thunderstorms. When a severe thunderstorm warning comes across the radio, social media or the TV screen, it tells us that thunderstorms capable of producing tornadoes, flash floods, destructive straight-line winds of more than 60 miles per hour and, finally, large falling chunks of ice known as hail, are observed or are imminent.

Many Western South Dakota's thunderstorms spew out hail. Most often hail the size of peas, golf balls or eggs plummet from the Western South Dakota sky. The most intense thunderstorms can even produce hail the size of softballs or even melons.

Most hail evolves from many up/down cycles in the thunderstorm cloud. In most areas of the world, hail formation begins in clouds adjacent to and usually upwind of the central cloud of the severe thunderstorm. These upwind clouds are called "feeder clouds."

Here's how it works. Drops of water in a cloud formation can remain liquid at temperatures as low as 40 degrees Fahrenheit below zero (equal to negative 40 degrees Celsius). That's correct; pure water can remain liquid, and not freeze solid, at temperatures as cold as negative 40 degrees Fahrenheit. These unstable "super-cooled" liquid cloud droplets freeze quickly when impregnated by tiny particles, such as dust, sand, silt, bacteria, factory emissions, etc. These now solid small ice particles form the hail "embryo."

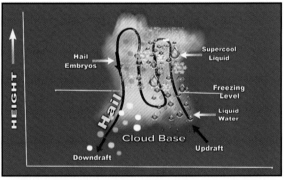

The life cycle of a hailstone

For a hailstone to grow, the downward traveling embryos located in a downdraft can and do reverse themselves when caught in the upward traveling warm, moist air column, loaded with cloud and rain droplets. The now upward traveling embryo in the central cloud becomes coated with a thin layer of water, which freezes after being lifted to the subfreezing portion of the cloud by the updraft. When the embryo again finds itself in the cloud's downdraft, the growing process repeats.

Three-inch hailstone, Black Hills, SD (NewsCenter1)

This up/down/up/down process can occur several times (depending on the strength of the updraft) before the hailstone grows large enough to overcome the updraft and fall out of the thunderstorm base due to gravity. Over time, this up/down/up/down travel of the embryo hailstone can result in a multiple layered, enormous, crop/structure/vehicle damaging hailstone. The stronger the updraft, the longer the hailstone develops, and the more significant the chunk of ice becomes before gravity takes hold of the situation and brings it to Earth.

During one of these upcoming summer days, about an hour or so past noon, a South Dakota farmer most likely will watch a singular thunderhead over the southwest sky go berserk and begin to spew out hail. Within minutes, that single thunderstorm can produce a swath of pelting hail snaking across a field of wheat, clipping it flat and absolutely level as if an enormous lawnmower moved across the land. While food growers suffer the most from crushing hailstones, let's not ignore the massive amount of loss and damage that occurs each year to livestock and personal property in all regions of the Plains states and Midwestern states.

The agricultural sector of the U.S. economy has the most to lose from hail. Hail not only restricts the number of crops produced every growing season; it also affects the quality of crop yields.

Severe Weather Watches Lacking for Hills

A few years back, I attended a severe weather workshop put on by the National Weather Service's (NWS) Storm Prediction Center in Norman, OK. These are the folks who issue severe-weather watches, not severe-weather warnings, over the contiguous United States.

To refresh your severe weather knowledge, the NWS has set up a series of steps to inform the public about the onset of a severe weather event. It works like "get ready — get set — go" at the start of a race.

The local NWS office issues an Area Forecast Discussion (AFD) describing the area's potential for severe weather. The NWS personnel and media meteorologists primarily use this initial "get ready" step for planning purposes. The next step, the "get set" step, means the NWS issued a severe weather watch. The watch informs the public that conditions appear favorable for severe weather, even though the severe weather has yet to occur. Finally, the "go" step, implies that the NWS issued the severe weather warning, informing the public about the occurrence or imminent occurrence of severe weather.

When it comes to notifying the public about severe weather, please understand a WATCH means "severe weather is possible," and a WARNING means "severe weather is occurring, radar indicated or imminent." Understand a WATCH means "to be prepared," and a WARNING means "to take shelter." Know, too, that only the National Weather Service (NWS) issues WATCHES and WARNINGS, not the media. Storm updates and reports from the NWS's network of trained storm spotters also are relayed to the media through the NWS. The media only communicates the WATCHES and WARNINGS to the public, never issues them.

The local offices of the NWS, here in the Black Hill's region it's the NWS office located in East Rapid City, issues all severe weather warnings. Severe weather watches, on the other hand, are not issued by the local NWS, but by the Storm Prediction Center located in Norman, OK.

A severe thunderstorm watch outlines an area where hail one-inch in diameter or larger and where to expect destructive thunderstorm winds to occur during a three- to six-hour period. A tornado watch includes the large hail and damaging wind threats, as well as the possibility of tornadoes.

Severe weather can occur at any time of the year, day or night. But, as the sun climbs higher in the sky allowing temperatures to heat up, May and June stand out as the peak times for booming thunderheads to unleash a menacing barrage of large hail, damaging winds, slashing downpours and, on occasion, tornadoes.

By way of a quick review, increasing warmth, coupled with the seasonal migration of the jet stream over the Great Plains, increases the threat of severe thunderstorms. Also, the Bermuda High pushes abundant amounts of Gulf of Mexico moisture northward in a monsoon-like surge into the Upper Plains, while mid-level dry air from the desert southwest enters the mix offering up mid-level evaporative cooling and causing even greater churning over the Black Hills region, followed by stronger winds aloft causing the developing thunderstorms to tilt or shear. The tilting of the storm allows the down-rushing rain to fall away from the powerful updraft, enabling the updraft to continue its formidable rising motion to feed the storm its energy bearing moisture. As a rule, when all these severe weather ingredients come together, powerful thunderstorms quickly gather high in the evening sky above the prairies and hills of Western South Dakota.

Expect the severe weather to happen most frequently here in the Black Hills region during the evening hours between 6 PM and 8 PM, when springtime heating exerts its maximum "boiling" effect on the atmosphere. Severe thunderstorms in the Black Hills typically produce large hail and damaging straight-line winds. Less often, some of the more potent storms yield tornadoes.

Supposedly, when favorable conditions develop for severe weather in the Black Hills and surrounding prairies, the Norman, OK-based Storm Prediction Center issues severe weather watches to place us on alert even though the severe weather has yet to occur.

In turn, when radar or a trained observer detects severe weather, the local NWS office will issue a severe weather warning. This is a simple, reasonable, and effective progression of events to get the severe weather word out to the public. First the "it-may-happen" watch, then the "it's-happening-now" warning.

Typical watches cover about 25,000 square miles, nearly half the size of Iowa. In an ideal world the "get ready — get set — go" progression of severe weather steps would fall neatly into place before a severe-weather event. Also, in an ideal world, the Cubs win the World Series more frequently than once every century, and teenagers get up bright and early during the summer break.

But alas, we live in a flawed world. Years of disappointment have left me with little expectation that the Cubs will make it to the World Series again anytime soon. I had three teenagers living at home where 'good morning' is a greeting generally spoken over lunch. And, I have seen too many severe-weather warnings issued by the local NWS office without any advance cautionary watch by the Oklahoma-based Storm Prediction Center.

The last item I thought I could do something about, so I boldly asked at the workshop — why so many severe-weather warnings in the Black Hills area without first issuing a watch? Their answer perplexed me. They said, and I paraphrase, that many severe thunderstorms affect only a small region for a short period, making watches impractical. Watches are issued primarily for areas where significant severe weather is possible, or the severe weather threat is expected to persist for many hours.

As I now understand, organized severe-weather systems and not the isolated weather events more common here in the Black Hills area necessitate issuing severe weather watches. Also, their phrase, "significant severe weather," mystified me. I like to reason that if the weather is severe, it's significant no matter if it's over the South Dakota prairies or downtown Minneapolis, MN.

Unfortunately, here in the Black Hills, the notable absence of the advanced cautionary watch remains the rule and not the exception. If my memory serves me correctly over the past few decades, the Storm Prediction Center issued much fewer severe weather watches, despite the many more severe thunderstorm warnings and tornado warnings issued by the local NWS office.

Because this problem shows no signs of improving, let me offer up a simple solution. I believe the responsibility for issuing severe weather watches for this area should no longer rest with the far-off Oklahoma-based Storm Prediction Center, but with our local NWS severe weather office. The local NWS office currently issues severe weather warnings, so why not watches as well? After all, who else knows West River weather better than those NWS meteorologists who live, work and observe the skies over the Black Hills?

Now that I have expressed my concerns to the Storm Prediction Center, possibly we'll see a few more severe-weather watches issued before the onset of severe thunderstorms or tornadoes. NOT!!

Gravity of Moving Water Heavy with Old Lessons

Among the many outdoor blessings, we can enjoy, the Mickelson Trail — a converted rail line roadbed running through the Black Hills — must rank near the top of anybody's list of attractions. The trail, completed in September 1998, winds for roughly 109 miles from Edgemont, SD to Deadwood, SD in the middle of the Black Hills.

All rail line builders sought an easy grade, requiring as little expensive alteration to the land as possible. Long ago, even ancient road builders knew that the path of least resistance ran naturally along streambeds. The Mickelson Trail was no exception.

Water — fluid and heavy — flows from top to bottom, propelled ceaselessly by gravity. With time along the way down, the moving water will push against its embankments, scour the river bottom or flood over its banks and alter the landscape it traverses.

When I bike the Mickelson Trail, I always try to go with the flow and follow the water currents. I know that the direction the water flows conveys to me the path of least resistance, in this case, pedaling my bike. With the water flowing in my direction, I know that gravity helps me move comfortably along.

Allowing gravity to do most of the work for me, I can focus on the total enjoyment of the hills, trees, sky, flowers, streams, wildlife, and nature's ambient sounds far removed from engines, sirens, or jackhammers. And, I do this with little distraction from aching muscles.

I ride through the Black Hills not so much for the physical exercise, but more for the exercise of the spirit. Along one stretch of the Mickelson Trail, north of Hill City, the Newton Fork twists and turns, providing the trail rider or walker with a steady aural and visual resonance of bubbling splashing water in the nearby Crooked Creek. Those sounds of moving water can soothe the mind, calm the spirit, and refresh the soul. If you seek nothing more than a body workout, let's keep that exercise in places like weight rooms, athletic clubs and swim centers.

I understand why so many people choose to live near running water. They can open their windows and, over a cup of morning coffee, enjoy nature's relaxing sounds of flowing waters.

Now, I am sorry to have to bring up an overlooked and underappreciated issue. Since we no longer carry a bucket of water from the stream to the house for that cup of coffee, as did our ancestors, we may no longer appreciate the weight of water. Now allow gravity to move tens of thousands of those buckets of water, I believe our ancestors developed a genuine appreciation of the tremendous force rapidly flowing water represents. Those powerful forces in the past proved deadly, and simply because it's been a while since we carried a bucket of water, will again prove fatal in the future.

A car driving through a flooded street in Rapid City, SD (NewsCenter1)

To judge water's force as our ancestors did, imagine lifting and carrying a bucket full of water uphill. Even if you are in good shape, you know that hauling a pail of water will require quite a bit of your energy.

A typical bucket of water is about half of a cubic foot in volume. On some very high-flow days, Rapid Creek flows through downtown Rapid City at 300 cubic feet of water per second (CFS), or even much greater flows. That means, during any given minute, 18,000 buckets of water or more move past a given point. Those numbers represent a tremendous amount of force to move or destroy almost anything in its way.

At these 300 CFS levels, Rapid Creek remains in its banks. However, Rapid Creek can unleash even more thousands upon thousands of buckets of water when it spills its banks under flood conditions, allowing it to easily wash aside human-made and natural obstructions in its ever-widening path. It was estimated that a peak flow of 50,000 CFS of rushing water swept through Rapid City during the June 9, 1972 flood.

We must regain some of the sanity our forebears had about moving water. When streambeds flooded in the past, our ancestors had enough sense to move to higher ground, because they respected the weight of water. That same force of gravity that pulls me so peacefully down the Mickelson Trail can quickly generate enough water force to move people, automobiles, and homes along a rather destructive and possibly deadly path.

If you live near a river or stream, it will flood. As Rick said in Casablanca, "Maybe not today, maybe not tomorrow, but someday ..." You must have a plan ready before the next flood watch or warning. When floodwaters rise rapidly, you will not have time to think it over. Most important: Do not procrastinate moving out of flooding areas and onto higher ground. Waiting, even for a few seconds, could mean being cut off from roads by rising water. And, finally, never drive through moving water on a flooded road.

On a bike heading downhill, gravity works with you. But, defy it, and gravity will beat you.

Dan Lutz

Our ability to more precisely measure and use the myriads of weather data now being amassed would greatly improve our ability to anticipate nature's next move far into the future.

🍃 MEASURING THE WEATHER

Before hitting the meteorological books in earnest at Penn State University, the Air Force sent me with 25 other new second lieutenants to Weather Observation School, at Chanute Air Force Base, Rantoul, Il.

Besides learning about the various weather measuring instruments, I was instructed in how to read the weather-related code, how to plot surface and upper air charts, how to record visual weather observations, and how to interpret the various states of the sky. The one most central message learned was to occasionally stop and just watch the weather.

Simply taking the time to watch the weather refashion itself from one moment to the next, causes one to appreciate the countless number of moving parts that goes into creating the wind, the clouds, the rain, the snow, the temperature or let's just say making the weather. Any attempt to forecast these rapidly changing mutations of weather, demands one to know and quantify precisely all the weather's initial conditions. Forecasting the weather for the next hour, day, week or year requires measuring the weather's now, as precisely as possible.

To forecast the weather precisely for any length of time, whether by the human or mechanical brain, requires a thorough grasp of the present ongoing weather happenings around the globe and throughout the troposphere. A formidable task that is currently not possible.

The National Weather Service forecaster, the military forecaster, even the seasoned farmer or rancher, or any weather enthusiast, must know with some precision of measurement, the current weather ingredients. Not knowing and recording the weather right now means not knowing weather's future.

A Forecast is Only as Good as its Beginning

The weather forecast digital computer must know the weather's "initial conditions" before cranking out a forecast based on those all-important rock-solid universal laws of physics. The military forecaster must correctly interpret the meaning of the recorded values from the various weather instruments scattered around the airfield before briefing the pilot on the weather at the time of takeoff. The farmer might look at the direction of blowing soil as he plows up the back forty before deciding when to plant. And, the weather enthusiast might watch for a specific cloud type over the distant horizon before deciding to go ahead with the family picnic.

The great minds from the past believed that by knowing the "right now" entirely and correctly, they could predict precisely the future and replicate the past accurately. Mathematical principles of mechanics, thermodynamics, and momentum can predict the weather into the future, and for that matter, replicate past meteorological conditions, if and only if the right now could be genuinely and thoroughly quantified.

So, why haven't weather forecasts been 100 percent accurate? Why not predict the weather for tomorrow, next week or next year without risk of failure? Why can these robust and powerful weather computers not anticipate the next flood, the next devastating drought nor the next hurricane? The answer may rest on the wings of a butterfly.

Now, we must resign ourselves to the fact that reasonably accurate forecasts only reside within periods not much longer than seven days. Any forecast beyond seven days should be considered the guidance of large-scale weather changes, such as temperatures turning colder or warmer, and weather conditions becoming wetter or drier.

The familiar adage "garbage in, garbage out" assuredly applies to weather forecasting. An accurate and detailed forecast product requires precisely measured meteorological parameters at the start of the forecast period. The precision of initial measurements of temperature, humidity, pressure and wind speed and direction dictates the accuracy of the weather forecast, especially when even forecasting beyond 72 hours.

Unfortunately, the present imperfect state of weather-measuring instruments introduces an ever-increasing degree of error to the computer model's generated forecast. The fundamental predictability of the Earth's atmosphere decomposes quickly to garbage-like status over more extended periods.

Snow depth measurement in Rapid City, SD

The "garbage in" includes weather measurements by instruments limited to one or two decimal point accuracy. This unacceptable level of precision currently found in thermometers, anemometers, barometers, and hydrometers assuredly starts the weather forecasting computer off in a slightly wrong direction, which quickly amplifies with time, into a less accurate forecast.

Our ignorance of what's happening weather-wise, top to bottom, around this giant globe where we live also adds to the extended range forecast decay. By not knowing what we do not know, like the impact of that butterfly's flapping wings on the environment, we add increasing error to the forecast mix.

To illustrate the importance of getting the forecast off to a good start, I'll use the infamous wave you see at football games. Theoretically, when one section of fans quickly and uniformly rises to their feet, a sequence of adjoining fans rises to their feet, simulating a wave that moves around the stadium.

However, if the initial group of fans lack the numbers (only a few rises: quantity) and lack the enthusiasm (those who rise to raise their arms high over the head: quality) the wave quickly deteriorates to one last guy standing up with arms raised high only to realize the embarrassing situation of being the lone emissary of the now-defunct short-lived wave. On the flip side, if an entire section of fans rises with much enthusiasm, the wave will maintain itself better and propagate much farther around the stadium. The more fans to start the wave with much enthusiasm, the longer the wave holds together as it travels around the stadium.

Like the wave at a football game, the quantity and quality of the measured weather input parameters that start the forecasting process are critical to maintaining the accuracy of the forecast as it progresses over time. Like anything we try to do, a lackluster beginning often makes for an unsuccessful finish.

Our ability on the weather front to anticipate nature's next move three or four days down the road will markedly improve once we attain better state-of-the-art measurement techniques. Weather measurements with accuracy beyond one or two decimals would go a long way to improve the accuracy of both short and long-range forecasts.

Without precise initial conditions, the accuracy of the weather forecast fades quickly in time and space. Even in today's world of super-fast, super-smart computers with lots and lots of memory, slightly inexact measurements of initial conditions limit even the most sophisticated weather computers to produce better forecasts.

Tiniest Variables Can Lead to Drastic Weather Changes

As discussed previously, a lackluster beginning often makes for an unsuccessful finish. Our ability to foretell better the weather's next move rests on obtaining precise measurements of the myriad of current weather elements found throughout the atmosphere, as well as the oceans around the globe. A monumental task but, nonetheless, a necessary task.

In 1960, meteorologist Edward Lorenz clearly demonstrated the profound impact of the initial conditions on the outcome of the weather forecast. He used a set of 12 equations to simulate a single weather event that he ran through a painfully slow computer by today's standards. It took Lorenz's computer many hours to step through the numerous calculations required to come up with the answer.

Days later, he replayed that same sequence. But this time, instead of starting at the beginning with the same initial conditions, he began in the middle of the sequence to save time. He entered the numbers copied from the printout to three decimal points (i.e., .512) or three decimals shy of what the computer had stored and used at that point in time (i.e., .512706). After resuming the computation progression, he left the computer to do its thing, albeit slowly.

To his surprise, the answer came out not just slightly different than the first time around, but widely different.

Lorenz's results showed that knowing precisely, within many decimal points, the goings-on of the weather right now, or the most precise initial conditions, dictate the precision of predicting the weather. Unfortunately, the limitations of our instruments to measure these initial conditions precisely will always result in the tiniest of errors at the beginning of the sequence. These smallest of inaccuracies will always result in forecasts to swell with unacceptable errors over relatively short periods.

Lorenz concluded that weather forecasts would never be right on the money because the precise measurement requirements of the initial conditions will never be accurate to the extent required. For example, try measuring the temperature, water content, and movements of a parcel of air 3,000 feet above the surface of the Earth to the nearest millionth decimal point. Then try measuring all fragments of air everywhere around the globe at nearly the same moment. That monumental feat is not possible with today's instrumentation.

The Butterfly Effect, a representative example of the chaos theory, illustrates the variations possible from the tiniest differences in the initial conditions. The Butterfly Effect says that a butterfly flapping its wings in South America could affect weather patterns in Rapid City, thousands of miles away. In other words, given enough time and space, the tiniest occurrence can produce unpredictable and sometimes drastic outcomes by triggering a series of initial insignificant events into increasingly significant events.

The long-distance influence of a butterfly's flapping wings on a tornado

Without knowing precisely, the impact of that butterfly on the weather forecast, or the influence of all the butterflies flapping around the globe, how is it possible to predict accurately future weather fluctuations. Does the impossibility of accuracy mean all is lost? As it turns out, we turn to the theory of chaos, which attempts to explain that even though highly variable results do occur in dynamic systems sensitive to their beginnings, some measure of order and predictability exist buried deep within.

Current levels of instrument error, interpretation error between points of measured parameters, and, oh yes, not knowing the real impact of that butterfly flapping around in South America, add up to weather forecast faux pas. But, that aside, we might find order in the chaos. And just maybe, knowing some order exists within the chaotic system of weather observation and measurement, further enrichment of the forecast may be possible.

There's Order Amid the Chaos

In the chaos theory, even the tiniest of errors in measuring initial conditions grow over time into substantial errors, drastically changing the long-term projected behavior of a system. Applying this concept to weather, our inability to measure weather parameters such as temperature, wind, pressure or relative humidity to a high level of accuracy, quickly decays the integrity of atmospheric forecast models to predict the weather. In other words, without a high degree of order measuring the weather's now, computer model forecast predictions become chaotic and drift farther away from reality with time.

Order in a messy bedroom (Google Images)

Fortunately, beneath the apparent chaotic behavior of the atmosphere, a sense of order and pattern lurks lessening the impact of inaccurate measurements of the weather's now. Let me clarify this noticeable conflict between the two opposing terms, chaos and order, with an analogy that many parents confront daily.

I believe it's safe to say that in many homes the teenager's dreaded messy bedroom sparks many an ongoing battle between the teen and the parents. As many parents gaze into their teenagers' bedrooms, their entire being immediately senses the true meaning of chaos. In the eyes of their son or daughter, they perceive nothing, but the blissful harmony that accompanies order.

A parent may see clothes, books, magazines and — for lack of a better word — stuff scattered about and intertwined among wires that are somehow attached to gadgets that produce audio or video entertainment. The kids, on the other hand, visualize an ordered pattern of their worldly goods that allow them to dress, do homework, recreate and relax within their own structured and predictable world. Order indeed exists deep in chaotic behavior.

An ordered pattern found among chaos (Google Images)

It wasn't until after the onset of the computer age that we discovered that momentary random behavior of the atmosphere obeys a deeper order bounded within a defined region. To demonstrate this concept, assign different colors to the three points of a triangle, and to two sides of a die. Randomly start anywhere within the triangle, throw the die and place a dot halfway from a randomly selected starting point toward the triangle point with the same color that shows up on the die's face. By using this dot as the next start point, repeat this apparent random process millions of times, and you will see an orderly pattern or design evolve made up of the millions of dots within the bounds of the triangle. So, with some digging, the order can be found in chaos.

To dig up some semblance of order from what appears to be a chaotic system requires repetitive and boring calculations that number in the millions. Computers work quite well at mindless repetition, making it an excellent tool to find order from chaotic, error-prone measurements of the right now. Without question, we live in a chaotic atmosphere, but thanks to today's supercomputers relentlessly searching out ordered direction within our atmosphere, better and possibly longer-range forecasts may not be totally out of reach.

As strange as this may seem, future weather forecasts might involve the chaos theory to identify fundamental structures upon which to expand and improve atmospheric computer modeling in the quest for better predictions of daily, monthly, annual and long-term weather and climate changes.

Let's revisit that supposedly chaotic bedroom, snow pants and boots here, snow skies over there and woolly mittens in that far corner. The ordered weather forecast buried in this teenager's chaos tells me that the winter season arrived, and the forecast calls for snow.

Nature Knows When Bad Weather is Brewing

Nature is its own weather service. Millennia before forecasts issued by weather service offices, newspapers, radio or television weathercasts, astute sky-watchers foretold the future by observing the signs often in plain view.

On this continent, Native Americans honed their instinct to co-exist with nature. They found that building shelter on the southeast side of a hill protected them from the prevailing northwest winter winds. Pioneer farmers across South Dakota used their practical knowledge of weather, and its seasons, to plant and harvest their crops. Some seasoned farmers still relied on their skills to inspect the horizon for clues to changing weather patterns, perhaps noting wind shifts to foretell when to ready their livestock for blizzard conditions and severe cold.

Some of the collective wisdom revealed through the ages of weather watching are part of the lexicon. Written in the New Testament, in the book of Matthew, Chapter 16, Verse 2 contains the passage: "When it is evening ye say, it will be fair weather, for the sky is red." Later, seafaring types added: "Red sky at night, sailor's delight. Red sky at morning, sailors take warning." With time, many of these weather gems became weather lore. Some yield credible and beneficial information. Still, others do not.

If you believe that a groundhog's shadow foretells the end of winter, then I want to sell you one of those rocks that lets you know it's raining by getting wet. Given the amount of attention we give Puxsutawney Phil, you would think that if it were possible to pin a microphone on him, he would be on television in Philadelphia forecasting the weather. My discussion here is about weather lore that makes some sense with some scientific support.

First, the atmosphere itself often shows its hand to the discerning observer, providing tips on impending weather changes. Simply watching cloud shapes and cloud base height provides clues to nature's next move. Generally, the loftier the cloud base above the ground, more pleasant the weather. That's because elevated cloud bases indicate both dryness of the air and moderate atmospheric pressure in the area below the cloud base, two qualities needed for fair weather.

Trees and plants can forecast the next rain event, as in the popular lore that says, "When leaves show their backs, it will rain." It's true because, as trees grow, their leaves align themselves according to the prevailing winds. So, when a storm wind occurs, typically not the prevailing wind, the leaves tussle backward showing their lighter backsides.

The animal world creates its own set of predictors. Farmers note that "A cow with its tail to the west, makes weather the best; a cow with its tail to the east, makes weather the least." Again, there's truth to it. Typically, a cow will graze with its tail to the wind allowing the animal to look occasionally downwind to spot predators. The cow knows full well that smell will betray an upwind invader. In many High Plains states, an east wind foretells rain, and a west wind predicts fair-weather; thus, the grazing animal's tail becomes a practical weather forecasting tool.

Even the insect world can help offer information about current and future conditions. Crickets remain as one of nature's best and most accurate thermometers. They chirp faster when sensing warm temperatures and slower when sensing cold temperatures. Because of the incredible accuracy of these little chirping insects, one can easily calculate the ambient air temperature. Simply count their chirps for 14 seconds, then add 40 to the number of chirps counted and, presto, you have the ambient air temperature.

Insects tend to bite more before a rain event. Higher humidity and lower pressure, two ingredients needed for rain, also leads to the release of body odors, a tempting target for hungry flies and mosquitoes. So, before you swat that next fly, remember that it may be trying to tell you something about the weather.

Cloud Predictors of State of Mind and Weather

I like to believe that the most commonly employed weather exercise remains just watching the sky. Here's an observation we all can make without an instrument and any theoretical background. Any one of us can immediately determine if the sky is clear or cloudy by a simple upward glance. And life's experience quickly tells us that small puffy clouds foretell fair weather, while big dark clouds portend an impending storm. Cloud types help us to be better informed about the near-term past, the present and short-term future of our local weather.

By recognizing the differences in cloud location, shape, and altitude, we can anticipate with some confidence the type of impending weather and how soon it will occur. Clouds manifest many of the dynamic and complex weather processes high in the atmosphere that continually evolve and dissolve over short periods. Clouds signal the onset of many upper air weather changes in near real time. Even without access to the data from sophisticated weather instruments, we can be better informed about impending weather changes merely by knowing cloud types and height.

Different cloud development processes, as well as their final configuration, predicts impending weather changes, whether becoming wet or dry, hot or cold. Small puffy cumulus clouds imply a stable atmosphere and fair weather. On the other hand, towering cumulus clouds signal an unstable atmosphere and possible thunderstorms. Cumulonimbus clouds forecast the onset of likely severe weather.

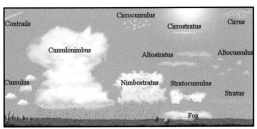
Troposphere occupying cloud types (Google Images)

Maybe I've been staring at the state of the sky too long because now I see some bizarre clouds that foretell my state of mind. For example, while sitting in traffic, "cirrus-congestus," clouds can be seen gathering overhead. If I stay in traffic too long, the ominous looking "migranous" cloud appears out of nowhere right before, and behind, my eyes. And finally, if road construction backs traffic more than two city blocks, a "cumuli-latte" with foam brews overhead. When that cloud drifts about, I rush over to the nearest coffee kiosk for a good cup of the notorious "sumtin sumtin" coffee blends.

Leaving the traffic in the rearview mirror, and while sipping the day's "cumuli-latte," my dog Wally and I can better enjoy our walk along the creek among the colors of the wildflowers and the sounds of the rushing water and birds. It's remarkable to note that after such a short walk in the park, how quickly those foreboding dark clouds lighten up to become "nimbo-stressless."

Raindrops and Bug Splats Gauge Weather

A few years back, after spending a few weeks traveling the Heartland, I took the opportunity to report on the Midwest's climatic conditions. Understand that these findings came about solely based on my observations and certainly without any rigorous scientific research. Therefore, no boilerplate, no preamble, no filler, only the bottom line. And, the bottom-line conclusion back then: It was wet in the Northern Plains and very dry in the Southern Plains. Given those same observations today, I feel quite confident I would come to the same conclusion.

This inference from the past came from first-hand observations of the landscape speeding by my bug-splattered windshield at about 75 miles per hour, give or take 5 miles per hour. My study region included only those High Plains states just west of the "jolly old baker," from the South Dakota grasslands, across Nebraska, Kansas, Oklahoma and into the Texas Hill Country.

South Dakota green grassland (South Dakota Department of Tourism)

The following states on the U.S. map form the "jolly old baker:" Minnesota (his hat), Iowa (his face), Missouri (his body), Arkansas (his pants) and Louisiana (his boots).

Jolly Old Baker

That year, these observations included a notably greener South Dakota landscape than those found in states to the south. Add to that fewer hay bales stacked in fields the farther south I drove.

Also, the counting-the-cows game saw numbers well into the hundreds across South Dakota and Nebraska, with much lower figures in Oklahoma and Texas. Apparently, those Southern states are selling off cattle in expectation of continuing dry weather. The Southern wildflower crop was wanting as well.

The Missouri and Platte rivers both appeared bank full. (My view of the Missouri river was from the now-closed Casey's Café in Chamberlain). Unfortunately, I cannot say the same for the Southern rivers. The Red River, which separates Oklahoma and Texas, appeared to be way below its banks. The Pedernales River in Texas, which meanders through former President Johnson's ranch and Willie Nelson's golf course, contained nothing more than a series of standing ponds linked together by sandbars. And, my all-time favorite river in Texas, the Llano, flowed mostly beneath its sand and granite rock bottom.

I have made this trip many times over the past half-century. In that time, I cannot remember not seeing a thunderstorm develop somewhere over Kansas or Oklahoma; if not directly over the highway stretching out in front of me, then certainly on the western horizon. No such thunderstorm happenings during that year's go around. Only in South Dakota did I enjoy watching thunderstorms grow, blow hard, rain out, and dissipate. I drove through some real brutes between Chamberlain and Rapid City.

There was nothing in the Kansas and Oklahoma sky except a bright sun shining through high, thin cirrus clouds. Regardless of the words sung by Three Dog Night: "I've never been to heaven, but I've been to Oklahoma." Both Oklahoma and Texas could use some rain to prevent them both from appearing more like a hot, dry hell than heaven.

And finally, the most visible indicator of the north/south change in macroclimate had to be the bug splotches on the windshield. I hypothesized that the wetter the weather, the higher number of bugs committing suicide on my windshield. In South Dakota, I had to clean my windshield every gas tank refill; in Texas and Oklahoma, every third refill. The unscientific through-the-windshield measurements back then suggested an extremely dry macroclimate, with possibly drought undertones, across the South, while the North enjoyed a typical to above-normal precipitation year.

Technological Eyes in the Sky Follow Weather

When doing the television weather, I usually discuss the past weather, followed by the current weather and then go on to explain why the current weather will change. The discussion about the past weather focuses on the satellite and radar image loops, the two most exciting and informative tools available to the weather forecaster.

Some weather satellites hover in a stationary position about 22,000 miles above the U.S., snapping pictures of the clouds, snow cover, smoke, volcanic dust, and blowing dust. Day and night, these satellites continuously document weather patterns as they evolve and move across the Earth far below.

Weather satellites collect two kinds of pictures: visible and infrared. The visible photos detect the sun's reflected light from cloud tops to show the cloud patterns and their location relative to the Earth below. Unfortunately, visible pictures of the clouds disappear into the darkness as night spreads across the U.S. Even though weather operates round the clock, light-sensitive photos work only during daylight hours.

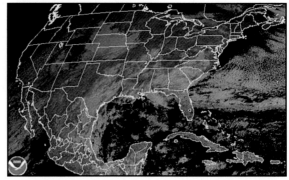

Infrared satellite map (National Oceanic and Atmospheric Administration)

When darkness arrives, and the weather continues in the night, infrared satellite imagery takes over the observation duties, to now play the lead role. Infrared photos allow for nighttime pictures by sensing heat instead of light, thus creating "temperature-sensitive pictures" of clouds.

Infrared pictures show temperature differences among clouds, providing the viewer with additional weather clues other than location and movement. For instance, the taller rain clouds have colder tops than lower clouds and appear on the infrared imagery as a brighter area than neighboring lower and warmer-top clouds. Severe thunderstorms reaching the top of the troposphere stand out greatest of all.

Infrared pictures make up most all satellite imagery shown on television, whether day or night. An added three-dimensional look introduced by weather computers to the two-dimensional images sent back to Earth transforms the image to something easier to understand.

Radar pictures display the past and current weather. When rain, hail, snow, sleet or freezing rain arrives anywhere in the country, including the Black Hills region, you can always count on a clear and colorful radar picture on television, Twitter, Facebook and your favorite weather app.

These pictures come about differently than satellite pictures. It works like this: the radar dish beams a pulse of energy into the sky and then waits for that parcel of energy to "echo" back. The amount of energy remaining measured in the returning radar beam, and how long it takes to get back to the radar dish, denotes the intensity and location of the storm, respectively.

Radar waves typically will pass through and not detect fair-weather clouds due to the small nature of the cloud droplets. When the bigger raindrops begin to form in the cloud, a portion of radar energy returns to the originating radar dish, indicating the location of the developing rain or ice. These echoes can tell the trained meteorologist when, where, and the size of the growing raindrops or hail found in the developing storm.

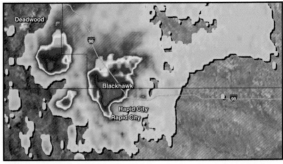

Radar map showing a thunderstorm over Blackhawk, SD (NewsCenter1)

Doppler radar adds yet another piece of valuable information to the radar mix. Doppler radar not only reveals to the meteorologist the intensity and location of the storm, and its embedded precipitation but also information about the air movement within the developing storm cloud. Knowledge about air circulation characteristics of the storm cloud can provide trip-wire information about tornado development 15 to 20 minutes before funnel clouds form and possibly tornadoes.

Both radar and satellite images provide a time-lapse history of storm movement. Having a comprehensive understanding of where our current weather came from, its speed, and direction of motion, and its intensity, allows the forecaster to quickly predict where the storm is going, when it will get there, and what to expect when it arrives.

Warm Up to the Idea of Change

Worldwide, the Earth continues to warm. Recent years' average temperatures tend to be warmer than in previous years. This warming trend concerns me — not so much about exceedingly warm past years, but because recent trends point toward even more warming in the years ahead. Some people believe that these warmer than normal years, both nationally and globally, indicate an aberration due to some freak culmination of errors. You see, instrument error, human error, and vast distances, in both space and time between weather-measuring locations, all play a role in bringing about the miscalculations in weather observations.

Historical trends give us the luxury of divulging insightful information by smoothing over these sorts of short-term observational errors. A historical study of all worldwide observations unmistakably suggests significant and worrisome warming of global climate over the past half-century. When

The changing climate (Yahoo Images)

combining global temperature over the past 25 years, the trend shows an alarming temperature increase in global warming per 100 years.

I now find it difficult to attribute the global warming detected over the past half-century to some anomalous episode somehow spiking out of the run-of-the mill sequence of regular weather events. Keep in mind that weather temperature anomalies do not infer climate temperature anomalies. If greenhouse gases that trap heat near the surface of the Earth — such as water vapor, carbon dioxide from the burning of fossil fuels, methane from landfills and coal mines, to name a few — continue to fill our atmosphere as they have in the past, history tells us the now-present global warming will continue with troubling results.

A few years back, I flew to Texas to visit my good friend, Don Larner, on his ranch near Dublin, TX. At 93 years old, Don continued to work his cows and, on occasion over a cold beer, he'll offer sound words of wisdom. He told me that the key to a fruitful life is the ability to adjust. "Change is inevitable, and if you cannot adjust to change, you will not persevere." As an example, Don loved to play golf, but when he could no longer swing a golf club, he adjusted his golf game, confining it to the putting green. Our global temperature continues to rise at an ever-increasing rate. Soon, it will be time for an adjustment either by nature or man. Nature's adjustment can be harsh — man's adjustment to limit climate-warming pollutants to the atmosphere can be manageable.

Science Says Climate Change is for Real

After years of debate and political haggling, the Environmental Protection Agency did an about-face by confirming global warming exists. After bringing together hundreds of years of comprehensive global temperature data, albeit flawed, the truth finally rises above the uncertainty. Proof of global warming can be found not only in climate records but in the secondary and tertiary repercussions of increased temperatures.

According to Gian-Reto Walther, an ecologist with the University of Hanover in Germany, warmer global temperatures, to date, have shifted crop growth, animal migration, and breeding. Plants in Europe flower and mature up to one week earlier. Birds now migrate and breed many days sooner. Red foxes and shrubs have migrated farther north into the Arctic. Once bare ground in the Antarctic is now covered with moss. Tree lines have climbed higher up mountain slopes, and butterflies have shifted their range northward. Oceans have warmed notably by 1.8 to 2.6 degrees Fahrenheit over the past 100 years.

The EPA believes this trend toward ever-rising temperatures most likely will continue well into the future, and hopefully, the world will adapt. Winters will be warmer, summers hotter, glaciers will melt away, shorelines will rise, trees will become more numerous and grain crops more plentiful. On the downside, more forest fires, floods, severe thunderstorms, more powerful hurricanes, and localized pestilence await us.

Can man predict the future outcome of things and make appropriate adjustments, like Charles Dickens' Ghost of Christmas Present, assured Scrooge that he can change the shadows of tomorrow, or is our future already set in concrete by the laws of nature?

Near the beginning of the 19th century, a French scientist named Marquis de Laplace argued that by knowing the complete state of the universe at any one moment, scientific laws then can entirely and correctly predict the future. Albert Einstein, too, proclaimed that "God does not play dice," implying all-inclusive laws that govern the universe determine its evolution, and not random chance.

Another German scientist, Werner Heisenberg, argued against this deterministic rationale. He demonstrated that, because it is not possible to measure the exact location and movement of an object, it is not possible to determine the correct and complete state of the universe and, therefore, man can never predict future events. Heisenberg argued that light must shine on the object to measure its correct position and speed, but once the light hits the object, the energy from the light changes the object's position and speed arbitrarily, and therefore, cannot be measured precisely.

I accept the uncertainty inherent in predicting the future using current laws and supercomputer weather models. I do favor the sound and proven mathematical science of statistics. Statistics uses past events to predict future events within established boundaries of acceptable confidence, and the statistical data clearly inform the vast majority of atmospheric scientists that global warming exists, and appropriate adjustment needs to begin now and not later.

History Suggests Dry Trend

Does a run of dry weather imply the beginning of a drought? It is difficult to answer that question because droughts, by their very nature, remain as nebulous beasts that never announce their arrival or their departure. At no time will the National Weather Service (NWS) report, "Today is the first day of a drought." Never will it say, "Today's rain officially ended the drought." No droughts had a distinct beginning or end. Only long after the adverse consequence of drought on our crops, prairies and water supplies, do we even acknowledge its existence? Only long after the rivers and streams return to bank full, and crops and grasses return to the fields and pastures, do we even acknowledge the drought's end.

Droughts defy simple labeling because its effects significantly impact our economy, our livelihood, and our well-being in entirely different ways. Weather conditions leading up to an agricultural drought differ from the beginnings of a municipal drought, which in no way corresponds to the precursors of a hydrologic drought, which do not resemble a recreational drought. So, the term "drought" takes on many roles and definitions.

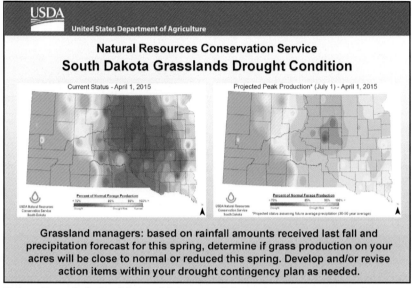

Palmer Drought Index (National Oceanic and Atmospheric Administration)

The Palmer Drought Severity Index (PDSI), widely used by many governmental agencies, including the NWS, attempts to categorize droughts under one umbrella — pardon the pun. The monthly PDSI not only uses rain and snow measurements but temperature, vegetation, and soil moisture information as well. The index smooths the data over short-duration wet or dry periods to lessen the effect of an abnormally wet month in the middle of a long-term drought. The data smoothing also reduces the impact of a few months of near-normal precipitation during a severe drought, which may falsely suggest the drought's end. Ideally, the concept of the PDSI attempts to be rigorous across the country, meaning a PDSI moderately dry period in South Dakota means the same as a PDSI moderately dry period in South Carolina.

The Black Hills' historical record of PDSI dates to 1895. A cursory study of these records indicates oppressive drought not being a long-term problem in the Black Hills region, except for the Dust Bowl years between 1934 and 1941. This same record also suggests that moderate, severe or extreme droughts in this region typically last about two to three years before ample rain or snow returns. On the other side of the ledger, periods of wet weather appear to last slightly longer, generally from two to four years before the next dry spell.

Up, Up and Away with Balloons

Every morning just before dawn, a single shadowy figure will slip quietly across its East Rapid City backyard, rushing toward a building looking every bit like a miniature celestial observatory. This mysterious figure makes the same trek twice a day at 12-hour intervals every single day of the year. With the distant lights of downtown Rapid City as its backdrop, the individual will not observe the stars and planets scattered across the black cosmos but will probe the very lowest part of the heavens, the troposphere.

The rawinsonde building located at the National Weather Service in Rapid City, SD (NewsCenter1)

Fortunately, many figures scattered across the country and, for that matter, around the globe, will dutifully perform the same early-morning mission at precisely the same moment. The early morning shadowy figure is a meteorologist with the Rapid City National Weather Service (NWS). And, like every other morning, the meteorologist releases one of literally thousands of weather balloons launched daily around the world.

In my opinion, that domed building high on the hill over the South Dakota School of Mines and Technology campus houses an essential weather instrument in the NWS's data-gathering arsenal. The meteorologist ties a small device called a "rawinsonde" (Radio WINd SONDEr), to a parachute, which is attached to a partially hydrogen-filled weather balloon.

Once released, the balloon and its data-gathering cargo drift upward, meandering about the airspace over West River. The balloon typically carries its cargo to altitudes of more than 100,000 feet or about 19 miles above the Earth.

As the balloon climbs higher into the troposphere, the outside air pressure decreases, allowing the balloon's girth to expand larger and larger. At some point high in

Rowinsonde in flight (Google Images)

the heavens, the balloon will burst, the parachute will open, and the rawinsonde will float back to Earth. When the balloon-borne rawinsonde begins its fall, its job is over.

The rawinsonde, consisting of pressure, temperature and humidity sensors, continually collects those weather data as it ascends through the troposphere, transmitting the information back to a receiving unit on the ground. Following the balloon's movement with precision tracking equipment determines the air's wind speed and direction, as well, at all levels of the troposphere.

The NWS transmits the now collected and recorded upper-air data to Suitland, Maryland, where powerful computers wait to gobble up well over 50,000 weather observations made every day. These weather measurements come from not only land-based balloon-borne rawinsondes located across the country but from additional upper-air observations collected from ships, airplanes, and satellites. The most advanced supercomputers ingest these data for the sole purpose of regurgitating weather forecasts.

For these supercomputers to predict what the weather will do tomorrow and days beyond, detailed and reliable upper-air data, collected by these balloon-borne instruments, becomes imperative. Little if any data contamination by friction or turbulence, due to Earth-bound topography or human-made structures, exists at the upper levels, thus making for smoother transitions of weather regimes into the near future. By inferring the upper-air wind, temperature, and pressure forecasts to surface changes, the forecaster gets a better understanding of the surface forecast. And, in the eyes of the public, it's at the surface where the weather forecast counts the most.

Often these weather balloon instrument packages launched from Rapid City, SD will most likely float back to the ground either over the Badlands toward Valentine, NE, or more east and northeast over the Philip, SD area. If you find one of these little white boxes lying on the ground, you can return it to the NWS for reuse. Approximately 30 percent of the instruments launched from Weather Offices in the U.S. fly again after recovery.

Santa Knows Weather

"If you see it in The [New York] Sun, it's so. Yes, Virginia, there is a Santa Claus." And you can find him here in Rapid City at the Rushmore Mall during the Christmas season.

Decades past, I sat on Santa's lap, wondering what gives with this bearded person. I later learned that the bearded person delivered gifts to all the good little girls and boys around the world. I did not care how he accomplished this mind-blowing feat, as long as he found my house, and I got my gift.

The author on Santa's lap many years ago (Marie Riggio)

Lucky for me a few years back, now a lot older, I was fortunate enough to spend a short time with Jolly Old Saint Nick and ask him some burning questions, including the how question. You see, Santa was kind enough to grant me an in-depth interview during a live news broadcast.

I also asked him a few questions about the weather's influence on delivering Christmas gifts to all the good little girls and boys around the world. Santa's answers to a few of my weather-related questions follow:

Bob: "How high do you and your reindeer normally fly on Christmas Eve?"

Santa: "We only fly at high altitudes when we are crossing large bodies of water like the Pacific Ocean. For example, when flying from Australia to Japan, we climb to about 30,000 feet above the ocean surface. The rest of the time is spent close to the ground, hopping from rooftop to rooftop."

Bob: "I am sure you know about the jet stream located high in the atmosphere. These winds are found within the much broader belt of winds that regularly blow from west to east around the world. Because the jet stream can generate wind speeds two to three times greater than the broad westerly winds, do you take advantage of these winds to expedite your Christmas Eve trip?"

Santa: "Over the past 362 years, I've become very familiar with this relatively narrow channel of strong winds. I have learned to avoid this area of strong wind because I always fly toward the setting sun to maximize the nighttime hours. Occasionally, I fly against the westerly winds, and it's imperative that I keep the sleigh away from the jet stream."

Bob: "Clear Air Turbulence or CAT is an invisible danger to aircraft because it can cause violent shaking. Abrupt changes in wind direction and/or wind speed within short distances causes CAT. CAT often strikes without warning, and it has been known to injure "unsecured" passengers on commercial airlines. Have you ever experienced CAT?"

Santa: "In the past, we have hit unexpected turbulence that shook the sleigh up a bit. And for that reason, I always keep my seatbelt buckled when in the sleigh. Secondly, we keep all the toys well secured, and I am happy to report that we have never lost a toy to air turbulence. Let me add that Rudolph has a nose for severe weather, including thunderstorms and ice storms. I find myself depending on him to guide the sleigh around dangerous weather conditions."

Bob: "Finally, my most burning question. How can you possibly visit every house in the world on a single night?"

Santa: "That's easy. I live at the North Pole for a good reason; from there, every direction is south. All the windows of my house face south. Look at a globe, and I think you'll get the picture. On Christmas Eve, I travel in every direction at once, simply by heading south. How quickly would your errands get done if they all lay in one direction? It makes a big difference. Secondly, and most importantly, the meridians, or those north/south lines that spread out from a single point at the North Pole, establish all time zones. Therefore, the North Pole is in every time zone, and in none of them. If I ever fall behind, all I need to do is go home and head south again. Everything I do has a simple and logical explanation."

So, there you have it. If you need any clarification of these questions, or if you have your own, stop by and visit Santa during the next Christmas season. He will most certainly have the answers.

Dan Lutz

Despite the local coffee shop banter, there have been notable improvements in weather forecasting accuracy since the 1950s, and more so over the past decade.

🍃 FORECASTING THE WEATHER

Despite the local coffee shop banter, there have been notable improvements in weather forecasting accuracy since the 1950s, and more so over the past decade. It's not perfect, but still, the public expects forecasts to be right on the money every time. I find no fault with that expectation because weather forecasts influence the everyday lives of most everybody 24 hours a day, seven days a week, 52 weeks a year, impacting their well-being, their work, and their play.

When severe weather is expected, the timely severe weather watches and warnings can and do save lives. Mid-range forecasts are used by industry to improve operational efficiency, by local governments to warn the public about unhealthy air pollution, and by the public to plan a wide range of daily activities.

Meteorologists strive to provide the best and most timely weather forecasts possible. To accomplish their mission to improve forecasting proficiency, they must possess a fundamental understanding of applied atmospheric processes, the proper use and interpretation of information provided by advanced technological tools, including high-speed computers, numerical forecast models, meteorological satellites, and weather radar.

Thanks to the rapid development of computers and numerical prediction models, the most significant advancements in weather forecast accuracy rest within the one-to-seven-day range. The American Meteorological Society (AMS) reports that today's accuracy of the two-day forecasts of sea level pressure remains as good as one-day forecasts a few decades ago. The AMS also states that the five-day forecast precision today is now as exact as the three-day forecast accuracy a decade ago.

With the latest advancements in computer hardware and numerical weather models, the public receives much better weather forecasts today than a few decades back. Even so, it still requires a rational thinking human to make the final forecast decision, based on experience, knowledge, and the intangible gut feeling. After all, where's the fun in blaming some unfeeling and uncaring piece of hardware for missing a forecast, when you can get more satisfaction blaming the human?

Weather Definitions Begin on the Front Lines

Meteorology, being a complex science subject to endless human interpretation, can at times express itself in abstract scientific terms unfamiliar to the layman, requiring the public to have a good deal of faith in things unseen.

When you watch a television weather segment or look at the local newspaper's printed weather maps, you often see several common symbols, each denoting the position of a specific weather phenomenon, such as cold front, warm

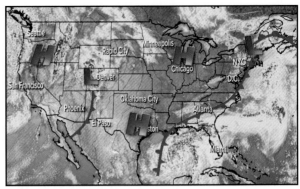

Common weather symbols on a United States weather map

front, low-pressure area, high-pressure area, trough, etc. But, just as visible markings on the ground denoting state lines or county boundaries do not exist, the same holds for those weather symbols. No symbols of these weather phenomena seen on the map exist as you peer out the window.

Meteorologists, broadcast meteorologists in particular, often assume the public fully comprehends the apparent elementary building blocks of weather logic, preferring to expend mental strain on the deeper abstract, creative interpretations of atmospheric conditions. Reasoning that they cannot thoroughly explain away the uncertainty of the weather forecast using basic map symbols, too often, broadcast meteorologists spend most of their allotment of time explaining the theoretical interpretations, instead of clarifying what they believe to be evident to the listening public.

So, not to assume all weather buffs understand rudimentary weather prediction, let us now go back to weather basics. By basics, I mean the easily explained what and why of the weather forecast, along with some clarification of those common symbols found on weather maps.

Previously, I discussed the Hs and Ls on the weather map. The blue H represents high-pressure systems on the weather map, which typically means clear skies, light winds, and dry weather. The red L represents low-pressure systems on the weather map, which customarily means cloudy skies, gusty winds, and wet weather. Also, various types of weather fronts show up on the weather map.

The term "front" came into widespread use by the weather community after World War I to better describe weather patterns. During that war, Americans and Europeans became familiar with the idea of troops capturing territory along battlefronts. People understood that soldiers huddled in trenches along a front represented the leading edge of territory claimed, lost or defended and not just a line. Often the front marked the location where the most violent clash of war took place.

Similarly, a moving weather front, either a warm front or a cold front depending on its direction of advance, represents the location where one air mass first overtakes an opposing air mass. If a mass of warm air overtakes a mass of colder air, the front is a warm front. We denote that on the weather map in red, with barbs showing the direction of travel, usually from south to north.

A cold front represents the leading edge of a cold air mass, replacing a warm air mass. We denote a cold front on the weather map in blue. If a cold and wet Pacific air mass heads for South Dakota, a cold front would depict the leading edge of that approaching cooler air mass, that most often changes the weather, temperature and sky conditions.

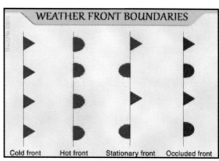
Weather fronts used on a weather map (Yahoo Images)

Bottom line, if you know in advance the timely movement (both direction and speed) of these symbols found on the weather map, you too could make a somewhat educated guess about what the weather future has in store for your outdoor activities. Will the temperature become colder or warmer, will the winds become stronger or not, and will the skies cloud over and rain or remain clear and stay dry?

The operative phrase here is "know in advance the timely movement." To fully understand the future actions of these basic symbols on the map, meteorologists must understand the theoretical interpretation and their application of many atmospheric scientific disciplines. It's the mathematical structure of these disciplines, such as the hydrodynamic and thermodynamic principles, that calculate the well-timed forecast movement of these weather representations across the weather map from today to tomorrow to the day after and beyond.

Forecasting Relies Upon Art and Science

Forecasting the weather endures as a blissful marriage of art and science. But, with every additional gigabit of data crammed into a computer's electronic memory by those technicians dressed in funny suits while working in a clean room, it becomes less art and more science.

During the early years of computers, the four essential ingredients in a good forecast recipe were: an understanding of fundamental atmospheric physical principles (clearly the science part), a whole lot of experience (the art part), a dash of luck (maybe a few dashes of luck), and a good dose of memory. Mix all these ingredients in a room with large windows to the outside, facing all four directions, and the forecaster of the '50s and '60s will be able to produce a weather forecast, albeit unhurriedly, with some degree of confidence.

Two of the four recipe's ingredients — basic science and luck — remain relatively unwavering from one year to the next. Since the universal laws of nature have withstood the test of time for eons, I like to think basic atmospheric science stands rock solid. Luck, by its random nature, also remains reasonably constant. Some days luck stays on your side, while on other days someone else walks away with the winnings. On average, a flip of the coin or a "push" best describes luck's contribution to the forecast.

Removing basic science and luck from the forecast recipe leaves two ingredients that have indeed changed with time: experience and memory. And, therein lies one of the primary reasons, forecasts have become more reliable over more extended periods. I'm not speaking of human experience and memory, but I'm talking about computer experience and memory.

Like humans, computer experience evolves with age; unlike humans, computer memory advances with age. I prove this dichotomy every year on my wife's birthday. I insist that her birthday is August 19, while she reminds me that the family will sing "Happy Birthday" on August 18, with or without me. Now I hide a cache of birthday cards around the house, so that, when August 18 comes around, and I wonder why the kids are setting-out a birthday cake, I will at least have a card handy to prove that I remembered this year's birthday celebration; a simple but effective arrangement I learned from many embarrassing memory lapses.

Today's supercomputers with multiple gigabits worth of remembrance translate into an increased experience and much more available memory storage. Computers gain experience and memory with time, while we humans may hold onto the expertise, but try as we may, find ourselves losing the memory with age to recall readily those prior bits of knowledge.

Computer memories increase over time, a distinct advantage in the weather forecasting game. Computers leave us humans in the dust when it comes to ingesting, storing, recalling, and acting upon billions of bits of continually changing data points collected all over the world. Big brain computers work over these recently received data points, as well as those pieces of information stored from previous periods, to produce one-to-seven-day forecasts and beyond all within an ever-shrinking computation time frame.

The always increasing gigabit of memory lets the computer gain more and more experience by allowing it to recognize better similar past weather events and use that historical information to produce much improved weather forecasts. The National Hurricane Center's hurricane and tropical storm predictions uses pass storm events to predict future movement. Their supercomputer recalls all past hurricane tracks and large-scale weather patterns surrounding each historic track at the time and uses that information to extrapolate future movement of ongoing hurricanes, tropical storms or depressions.

Humans must still play a role in the weather forecast business. Even with our ever-increasing dependence on smarter and faster computers, the weather remains an inexact science. The imprecise happenings of weather leaves an opportunity for most meteorologists to argue the case that on occasion birthdates may slip our memories, a real-live thinking person with all our human frailties must still interpret the many available computer products and choose the one forecast product making the best sense for the current weather situation.

Now, if I could just remember where I hid those birthday cards!

Weather Watchers Hang Out with Models

Back in the days of typewriters and slide rules, weather forecasts began with a long and thoughtful gaze out the window. Today, it starts with computer-generated weather-simulation mathematical models. The basic concept behind these complex and rigorous models compares nicely to dropping a stone in a pond of water, creating concentric ripples flowing outward from the initial splash. By measuring the exact weight and dimensions of the stone and authenticating specific attributes of the pond and applying this information to the proven laws of fluid dynamics and physics, computer-generated mathematical calculations will tell us the exact speed and intensity of the ripples at any time and location on the pond.

The same concept holds with weather-simulation computer models. Ripples of weather can be simulated at any time and location across the country by carefully measuring current weather conditions, understanding the physical characteristics of the surrounding landscape and applying this information to the fundamental laws of thermodynamics and atmospheric physics.

In truth, sad to say, these computer models can never be completely accurate in forecasting the weather. Their imperfection, as discussed previously, arises from not knowing with an extreme degree of accuracy, the measured initial weather conditions. Problems also evolve from the many assumptions about interpolations and extrapolations of weather data measurements across both time and space.

One way to lessen the effects of these imperfections is to use bits of information gleaned from many different computer models. The assortment of computer models available to forecasters over the years include names with acronyms such as HRRR (High-Resolution Rapid Refresh), RUC (Rapid Update Cycle), NGM (Nested Grid Model), ETA, AVN (Aviation Model), MRF (Medium Range Forecast Model), NOGAPS, ECMWF, UKMET, MM5 (Mesoscale Model 5th generation) and so on. Except for the HRRR and RUC, these computer models generate short- and long-range weather forecasts for further interpretation twice a day.

Some of the more current and advanced computer models now include: RAMS (Rapid Atmospheric Modeling Systems), RAP (Rapid Refresh Model) which runs hourly, replacing the RUC; the NAM (North American Model) which runs four times a day replacing the ETA; the

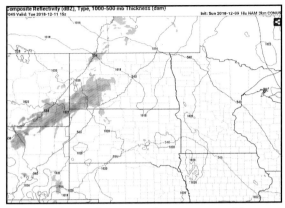
An output forecast from a weather computer model (Google Images)

GFS (Global Forecast System) which also runs four times a day, replacing the AVN and the MRF; and finally the WRF (Weather Research and Forecasting Model), replacing the MM5. The HRRR computer atmospheric model updates hourly providing much-improved cloud and convective storm forecasts.

There is no such thing as one forecast model that fits all forecasts. Each forecast model has its strengths and weaknesses. For instance, the RAP model runs every hour, making it readily available for updates using the most recent observations; however, it can only effectively forecast the weather out 36 hours. The NAM model provides a more detailed forecast across the country by utilizing a higher resolution of weather parameters and a terrain-following coordinate system. The NAM model works well in forecasting amounts of precipitation, mainly winter snow, in mountainous areas; however, this model tends to generate widely different ripples of weather change from one run to the next.

Because of the superior resolution of data points, the RAP and NAM models perform best for short-term forecasts. The models with less resolution between data points, such as the NGM and GFS models, tend to do better with predictions beyond a few days. The NOGAPS and the ECMWF models do best predicting seven-day outlooks. RAMS uses a fine mesh grid to simulate localized atmospheric systems such as thunderstorms, while simultaneously modeling the large scale environment of the systems on a coarser grid.

The dynamic nature of weather often causes the output of these many models to differ notably among themselves over even a short time frame. The key to forecasting the weather here in the Northern Plains or anywhere in the country rests solely with deciding which of the models' output best represent the current atmospheric ongoing trends, the forecasters gut feeling about tomorrow's weather aided by years' of experience, and finally, by merely glancing out the window.

Computers Crunch Weather Models

When discussing computers, you either speak of hardware or software. Hardware is just that: if hardware falls and hits you on the head, it hurts. If software — a set of written program instructions that tells the computer's brain what to do — hits you on the head, it's just a minor annoyance.

The electrons in today's supercomputers do a much better job of number crunching and recalling historical details than the electrical impulses in our brains. Computers record those electric charges as a series of on-or-off, all-or-nothing, yes-or-no recollections. As time and technology progress over the years, computer memory chips get smaller, but pack more memory capacity into the ever-shrinking space, leading to smarter, and smaller computers. The faster the computer (hardware), the better it can digest and utilize the instructions (software) quickly and precisely. Placing full faith in past technological trends, expect even faster, smarter, and more compact computers in the years ahead.

Now that today's computer chips make available enormous amounts of capacity for memory, they can quickly crunch the extraordinarily complicated mathematical equations found in today's weather forecast models. These mathematical weather models now efficiently and rapidly simulate pressure, temperature, humidity, and wind values that make up our atmosphere. With today's computers at our ready, forecasting short-term and extended weather conditions quickly and more accurately most certainly requires more science than art.

Forecast computer models evolved over many years of research at institutions such as Penn State University, Colorado State University, The University of Wisconsin and the South Dakota School of Mines and Technology (SDSM&T).

Two of the early pioneers in the field of cloud modeling did most of their research at the South Dakota School of Mines and Technology (SDSM&T) back in the late '60s and early '70s. John Hirsch began using computers to study cumulus cloud development over the Great Plains. Dr. Harold Orville introduced more sophisticated cloud physics, surface terrain characteristics, and time steps to study cloud development over the Black Hills.

The addition of cloud physics caused Dr. Orville's cloud model to simulate rain or snow. The introduction of time steps allowed the computer cloud model to replicate the birth, growth, and decay of mountain-induced rain/snowmaking cumulus clouds.

The research and development of computer cloud models conducted at the SDSM&T may appear insignificant when compared to the software making up the global forecast models available for today's supercomputers, but the accomplishments and lessons learned were explicit building blocks of knowledge that continue to play a significant role in today's weather forecast models.

The final scene's monologue from the 1957 classic science fiction movie, "The Incredible Shrinking Man," explains the fallacy in the human tendency to disregard the universal scheme of minimizing the initial building blocks from past events that led up to today's high-tech state-of-the-art products. No matter how small each component of research happens to be, each part remains vital to the final product. As the last line in the movie states, "To God there is no zero. I still exist."

Weather: The Disorder of Things

The root of weather forecasts lies in the chaos theory or the second law of thermodynamics, which states that as disorder becomes more prevalent with time, it releases energy. The second law of thermodynamics also tells us that energy flows from hot to cold, resulting in warm. Without outside sources of energy, it cannot flow from warm back to hot. The weather forecast represents nothing more than a prediction of weather events changing from order to disorder.

On any given spring afternoon, the skies will remain a deep blue high above the quiet atmosphere, as long as the atmospheric energy, usually in the form of water vapor, stays concentrated, ordered and uninterrupted. All bets are off when weather conditions change to allow the release of the water-vapor-stored-energy as the now peaceful atmosphere transitions from order to disorder.

Order and chaos (Google Images)

A valuable and applicable weather forecast predicts when an atmospheric trigger comes along to change this tranquil scene from order to disorder/chaos. A quick and spontaneous energy release initiated by a trigger of sorts can mean colder temperatures, rain/snow, and strong winds across the state. Thunderstorms, lightning, wind, tornadoes, or hurricanes (all the sensational components of weather) also happen because of that second law of thermodynamics forcing disorder from order.

Recall that according to the second law of thermodynamics, energy spontaneously flows only from concentration (order) to diffusion (disorder). Everything, in this ordinary world of actual objects, will eventually wear out, moving from order to disorder, as we too demonstrate a similar effect growing older. Like it or not, our bodies slowly give off energy as we age. Leave any human-made structure unattended, and over a period it will deteriorate and collapse, as it too slowly releases energy.

There is a tremendous amount of energy locked away in a forest of trees. That energy will be released, either slowly through age and decay, or rapidly, through fire. A single spark can act as a trigger, starting a forest fire and demonstrating dramatically a rapid release of stored energy liberated over a short time. Releasing energy rapidly by a trigger, or slowly over decades, gives off the same amount of energy. Only the time required to accomplish the energy purge varies.

When forecasting the weather for the next few hours or days, forecasters try to identify those meteorological triggers embedded in the current observations or from the various computer models described earlier, that may start a rapid release of energy in the heavens above the mountains, oceans, plains, and forests. Identifying the trigger early on provides the forecaster enough time to issue timely warnings of such near-term influential weather events as winter storms, thunderstorms, flash floods, hail, hurricane-force winds, and tornadoes.

Just as a tiny match can trigger a massive forest fire, there are triggers in the atmosphere that can start the quick release of raw power stored within the chemical bonds of water vapor. Common triggers to kick start weather include fronts sliding under a humid and unstable air mass, upper-level pockets of cold air causing strong vertical instability, areas aloft where the airflow diverges, initiating strong updrafts that soon become thunderstorms and hot afternoon temperatures that give rise to clouds and rain showers.

Based on the latest forecast, we anticipate a pleasant summer afternoon to indulge in golf, long walks, or fishing in our lakes and streams without the threat of sudden severe weather. Then a trigger comes out of nowhere sliding unnoticed in between multiple data gathering stations to radically and quickly change the once pleasant afternoon into a chaotic evening of gusty winds, heavy rain, lightning and thunder.

Often these triggers slip through our widely scattered network of weather measuring instruments without detection. Large holes in our data measuring network exist in both space and time. Approximately 200 miles separate many of the upper-level measurement stations, for example, that only operate every 12 hours. This large gap in weather gathering instruments, both time and space, requiring interpolation between data points and extrapolation of data into areas void of any weather data, account for yet another reason why missed forecasts happen.

So, there you have it, the weather rendition of the second law of thermodynamics. I hope this gives you some understanding of why meteorologist occasionally miss a forecast. Don't blame the weatherman. Blame those spontaneous, yet sneaky triggers.

Weather Forecasts Remain an Imperfect Science

Forecasting the weather is not an exact science; not now nor anytime soon. Instead, the weather remains a complex, mysterious, and always changing phenomenon. A marvel that remains unpredictable with any certainty to this day. The uncertainty principle, which states that measurements can never be exact, tells us weather forecasts can never be right on target, no matter how hard we try. Also, weather forecasts can never be entirely correct in the eyes of the public because no one forecast can account for or tailored to everyone's own personal space and schedule. If the forecast called for rain and yet it did not rain in your flower garden, then yes it busted, but not so according to your neighbor whose garden was soaked by rain.

More times than not, the weather forecast will not address precisely the conditions within your sphere of reality because at least hundreds of thousands of different micro-climates make up the tapestry of our weather. Within areas as small as city blocks, changes in elevation, sun angle, vegetation, and ground make-up create a nearly endless number of different micro-climates of temperature, wind, and moisture. For example, south-facing slopes will typically be warmer than north-facing slopes, particularly in the Northern Plains. Colder air will flow downhill, creating colder overnight low temperatures in canyons and valleys. Homes with slightly lower elevations generally experiencing colder overnight low temperatures than their neighbors' higher-elevated homes. No one forecast can account for all people's weather requirements all the time.

Some meteorologists argue that forecast uncertainty should be conveyed to the user during each weather forecast and especially in the longer-range outlooks. Let's say the seven-day forecast calls for a temperature of 61 degrees on Saturday. Indeed, given the vast "weatherscape" here in the Black Hills and the always-changing nature of the atmosphere, 61 degrees will most likely be incorrect in many backyards. So, why not include: plus, or minus 10 degrees to the seven-day forecast to account for almost everyone's backyard?

Others argue that after years of forecasting improvement, conveying uncertainty will be a step backward. Despite the local coffee-shop banter about yesterday's busted record high-temperature forecast, today's two-day forecast, for a pre-determined location, is as accurate as the one-day forecast was a decade ago. A five-day forecast now is as accurate as the three-day forecast was in years past.

I believe the user and reader understand the inherent uncertainty of weather forecasts. In my opinion, adding risk to the weather forecast mix will create even more confusion and, yes, more doubt. No matter how much uncertainty forecasters try to convey, I believe the public wants to hear, "Yes, it will rain" or "No, it will not rain." To communicate a "yes/no" forecast effectively would help the public make better weather-related decisions; unfortunately, it currently remains one of the most challenging parts of providing useful weather information to the public, not only here in the Black Hills, but anywhere.

Seat-Of-The-Pants Flying and Forecasting

Some years ago, I enjoyed a delightful afternoon, high on a hill just east of Rapid City. This lofty position provided me with a magnificent view of the Black Hills silhouetted across Western South Dakota's deep blue sky. I, along with 35,000 other spectators, spent the afternoon enjoying the Dakota Thunder Air Show at Ellsworth Air Force Base.

Wing Commander Col. Przybyslawski (I like to call him Tony because I am informal and this way, I can avoid the embarrassment of the name pronunciation dilemma that all new base arrivals must go through), air show director Maj. David Bucknall and the men and women of Ellsworth AFB pulled off a delightful air show. It demonstrated the impressive skills of our country's military and civilian pilots.

My task at the show was two-fold. First and foremost, I had to keep Mother Nature's thunderstorms as far away from the air show's Dakota Thunder as possible. You see, thunderstorms and expensive aircraft do not mix and must stay far apart. The only thunder pounding the spectator's eardrums should be the booming sounds of aircraft engines.

The Thunderbirds about to take off at Ellsworth AFB, SD

Secondly, I had to talk about the show — the strengths of the different performing aircraft, and how those strengths are highlighted through the twists, turns and spinning aerobatics of the aircraft.

I can talk about the weather. I cannot talk about aircraft. I do know airplanes fall into two categories, big and small, and within each of those two categories, you have fast and slow. The B-1B bomber, for example, is a big and fast aircraft. I felt entirely inadequate on that grandstand.

So, how did I make it through this announcing gig in front of all those people lined three and four deep along the flight line? Easy, Lt. Col. Mark Morgan shared the microphone with me. Back then, Mark wore many hats at Ellsworth AFB, not the least of which was B-1B instructor pilot. So, here's a guy who can talk airplanes from the get-go to the get-stop.

His in-depth and complete answers to my simple questions made me look almost insightful and clear-sighted. The pride and love of aviation, as exemplified through Mark's explanations of the overhead and static aircraft displays, clearly expressed the genuine feelings of pride felt among all the airmen I associated with while at the air show.

Mark explained that the more sophisticated aircraft depends on a higher degree in computer technology but make no mistake; the pilot remains the ultimate computer. And with a smile, he went on to explain that seat-of-your-pants flying is not dead yet.

The same applies to weather forecasting. Computer technology now takes a lot of the art of weather forecasting out of the hands of the forecaster. But make no mistake, gut feeling, seat-of-your-pants forecasting, if you will, is not dead either.

By way of illustration, let's get back to my first task of keeping all thunderstorms at bay. By late morning, a massive thunderstorm developed over the Black Hills directly west of Ellsworth. The question Mark asked me over the public address system was, "Bob, will it move over the base?" Well, if some of those pilots could fly by the seat of their pants, I could forecast the weather by the seat of my pants. My answer, again over the public address system, "A guaranteed no."

No computers, no radar to rely on, just my gut feeling, and the direction of the anvil blow off suggesting the storm will move southeast and stay south of the base. And indeed, it remained far to the south, allowing the air show to go on without a hitch. Lucky me.

Numbers are the Language of the Sciences

Most psychologists will tell you to please communicate when trying to develop a strong and lasting relationship, whether between husband and wife, parent and child or even between me and my dog, Wally. To develop a better understanding between two parties, always express yourself clearly, and listen intently. This sound and fundamental advice carries into the sciences, as well, and — more specifically — into meteorology.

A better understanding of nature, through improved communication, places the forecaster in a more advantageous position to predict what nature will do in the near term. This now begs the question of how do we and our natural surroundings interact? The answer: by numbers; we communicate with our environment with numerals.

Numbers are the language of the sciences. Numbers are the universal language among all the peoples of our Earth, and probably any other inhabitants of this and other universes. Although we speak different vocal languages, enjoy different cultures, and operate under various governments, we agree on the language of numbers. For example, we recognize that the number one is always bigger than the number zero and smaller than the number two. All computers, both national and international, communicate through the number zero and one.

Nature uses a rather elaborate numerical language to communicate with us. Nature's language uses "real" numbers (i.e., 1, 6.7, -12.56), "imaginary" numbers (numbers that when squared, equal a negative number), and complex numbers (combinations of real and imaginary numbers).

At times nature talks to us through numbers without any dimension. These are quantities that stand alone without much meaning. To explain: 32 is a number without dimension. We know how large it is, but we don't know what it represents. But, when you specify 32 degrees Fahrenheit, then the number has some sort of meaning about temperature, by assigning it the dimension of degrees.

The Froude number is a number without dimension. This number allows the Black Hills to tell us if we can expect upslope snows. The Rapid City National

Weather Service (NWS) web page explains that the Froude number is, "A non-dimensional value based on a combination of the mean wind speed, the height of the obstacle, and the buoyancy."

The wind speed number tells us how fast the wind blows at predetermined elevations; the elevated Black Hills obstructs the wind; and, buoyancy determines the rapidity of rising and falling air. The Froude number assimilates all these multiple parameters into a comprehensible dialect the meteorologists at the Rapid City NWS can understand.

A Froude number less than one implies the winds will blow around – rather than over — the Black Hills of South Dakota. This non-dimensional number conveys to the forecaster to not expect upslope winds. When the Froude number is higher than one, upslope flow over the Hills may occur. And, given enough moisture in the air, upslope rain or snow may dampen the windward slopes.

So, here we have nature communicating with us by this single number and telling us whether upslope snows will or will not occur when the wind blows over or around the Black Hills. So now we know that nature speaks through numbers to let us know who will and who will not have to shovel snow.

Keeping Weather Real

Spotty precipitation across the Black Hills region manifests itself, particularly during the summer. The more widespread winter and early spring snows and rains change into the more splash-and-dash, spotty rain shower variety during the late spring and summer months. These scattered showers leave some areas on the prairie drenched with rain and the other regions badly in need of moisture.

A weather forecast cannot possibly address all the micro-climates that exist here in the Black Hills region or any place across the country. Therefore, any single forecast may be wrong at one given location but correct at someplace else close by.

I prefer to focus my forecasts on the airport's weather observations, partly due to tradition and partly due to the length of record found there. Initially, airports or military air bases accommodated all weather offices, where current and proximate weather observations provided valued information regarding the safety of air operations.

Today, remote sensing allows some weather offices, such as our own Rapid City office, to move some distance away from the airport, all the while maintaining their weather observation instruments at the airport. In Rapid City, like most cities in the country, remote sensing instruments – no longer the human eye — measure the airport's temperature, dew point, pressure, humidity, wind speed and direction, visibility, weather type (i.e., rain, fog, snow, haze, etc.) and sky condition below 10,000 feet, at least once every hour. The East

Rapid City National Weather Service (NWS) office receives these weather observations instantaneously, to be verified, logged, and reported to the public as the official airport weather observation.

One glaring problem with this remote weather data setup has to do with cloud cover. If overcast clouds, such as altostratus, exist above 10,000 feet at the airport but clear skies below, the official observation reported is clear skies. Yes, clear skies below 10,000 feet but unfortunately not clear skies at or above 10,000 feet.

When I served as a forecaster in the U.S. Air Force, a weather observer was on-station 24/7 out near the runway, measuring and documenting weather observations at least every hour and often in between the hour depending on changing weather conditions.

The automatic weather station located in Rapid City, SD

When making my forecast, it helped to get the opinion from the weather observer regarding what he saw outside his viewing station. With the advent of remote sensing weather measuring instruments, it's not that way anymore.

In my opinion, a weather office cannot be a weather office without a thermometer, barometer, anemometer and rain gauge located somewhere on or near the same premises and not tens of miles down the road. These nearby instruments allow the meteorologist the opportunity to go outside from time to time to check and record the instrument measurements and, in doing so, feel, taste, and smell the weather. It keeps weather real to those people who predict it.

The Rapid City NWS office reflects my sentiment nicely. It, too, keeps weather measuring instruments in its backyard, as well as receiving automated weather observations from the airport. So, now you have a choice. If the weather observation at the airport does not fit your needs, try the NWS weather office weather observation. If neither, I'm sure you can find it somewhere in between.

Drought Causes Explained

The biggest ticketed and one of the most damaging events mother nature can throw our way is drought. Those ingredients that bring about drought go beyond localized weather changes and into the realm of large-scale climatic changes. It may take both terrestrial and extraterrestrial influences to effectively disrupt the sensible balance between abundant precipitation and insufficient precipitation for extended periods.

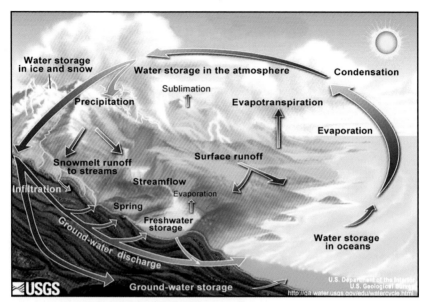

The water cycle (United States Geological Society)

Water covers more than two-thirds of the Earth's surface. As the wind passes over the vast oceans, it continuously takes up some of that water through evaporation. The air-borne moisture condenses to form clouds, which in turn allows some of the moisture to fall over land as fresh-water rain or snow. From there, it flows back into the oceans where the water cycle repeats itself. This circular series of events is known as the "hydrologic cycle" or the "water cycle."

Precipitation does not fall equally across the land. On average, some areas receive more rainfall than others. The scarce end of the water cycle produces desert areas, while the bountiful end of the water cycle produces rain forests.

If this global precipitation pattern remains consistent, life adjusts nicely and relatively comfortably to either a shortage of precipitation or copious amounts of moisture. It's when the disruption of the moisture cycle for an extended period that trouble starts in desert areas, rain forest areas or those areas in between, such as South Dakota.

Desert regions can and do expand and contract, due to long-term and subtle changes to upper-air patterns, which especially disrupts long-term rainfall patterns in those areas along the fringes. Some years, even decades, those fringe areas find themselves engulfed in drought conditions, while awash in moisture during other periods, only due to location. Fortunately, South Dakota finds itself unaccustomed to the long and harmful dry spells that accompany an expanding desert region, due to its great distance from those arid regions.

South Dakota droughts may result from sporadic sun behavior impacting ocean surface temperatures, which in turn may cause large blocking high-pressure systems to settle in over the Upper Plains. Since the sun stands alone as the sole source of energy for our atmosphere, any quirks or hiccups in its behavior, most assuredly affect the long-term weather patterns around the globe in some fashion.

Some scientists believe increased sunspot activity matches well with increase drought occurrence for some locations on Earth. As far as I know, no proven links connecting sunspots to drought exist to date. But, leaving sunspot activity out of the equation and just working a bit closer to home, linking ocean water temperature to drought indeed shows much promise to the burning question: why drought? Once knowing the "why drought" question, doors may open a bit more to lowering droughts' impact on people's livelihood by fashioning more time for drought planning.

The warming of the waters off the Peruvian coast initiates a weather event now labeled "El Niño," which, in turn, alters the polar jet stream causing a vast high-pressure cell to drift over the Midwest from either the Atlantic or Pacific Oceans. These El Niño-displaced large-scale entrenched weather systems produce many days of rainless and searing hot weather over South Dakota.

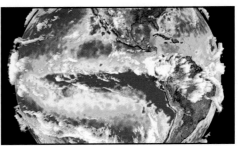

An El Niño event showing warm waters off the Peruvian coast (Google Images)

The drought of the '30s was due to the dominance of one such high-pressure cell. Storm systems migrating out of the Northwest Pacific Ocean with the potential to bring rain to the Midwest were shunted eastward across Southern Canada. Other places than the Midwest, benefited from these storms' potential for manufacturing rainfall.

When the next drought strikes South Dakota, its initial cause may be sunspots, which in turn, warm the ocean waters found off the South American coast. But the ultimate reason for the extensive chain of weather-changing events most assuredly will be a massive upper-level dome of high pressure firmly locked in place over the Northern Plains.

Drought: It's About Timing

Over the years, many conversations, particularly in the agriculture communities, centered around this distinctly insidious weather phenomenon called drought. Headlines read, "Region hit hard by killer drought," followed by reports about farmers and ranchers selling off cattle to conserve what little grain or usable pastureland remained.

Here's the rub, because the beginning and the ending of droughts remain so nebulous, it's long after we witness the characteristic drought signatures — parched pasture, withering crops, plummeting water supplies, and in more extreme cases, famine and skeletal remains of animals — that we can say with any certainty that a drought occurred. Typically, the official drought declaration occurs long after it's onset.

A dry brown hillside, Theodore Roosevelt National Park, ND

Will a two- to three-inch rain turning the brown landscape quickly back to a lush green, replenishing stock ponds and wells, and start the flow of now dry creek beds signify drought's end? Can we then say that that rainy day ended the drought? How about after a week or two of above average temperatures under cloudless skies, can we say a drought started? I think not.

If we never read or heard the news, those of us living in Rapid City or the surrounding Black Hills communities would have little clue of an ongoing agriculture drought. Hey, Rapid Creek flows through town like always, the golf courses are a deep green, swimming pools are brimming with water, and there is plenty of water coming from our faucets to satisfy our thirsty selves, garden, or lawns.

How can there be a drought when our recreation activities in the Black Hills are not impaired? Pactola, Sheridan and Angostura reservoirs all have plenty of water for boating, fishing, and swimming. I'm no biologist or park ranger, but the normal fish and wildlife propagations appear normal without any adverse drought-related impacts. Also, the tourism industry appears to be doing well during the spring and summer seasons.

So again, I ask the question – Can a few dry springs and early summers imply we are in the throes of a drought so deserving of our attention and concern? I believe so. You see, we live in an agricultural state, and drought first shows its ugly head in that agrarian pursuit. Even though an agricultural drought's tentacles have yet to impact our municipal water needs, recreational and tourist economy adversely, like the prominent droughts in the 1930s and 1960s, a few dry planting seasons most certainly can have a damaging impact on farming and ranching.

Try to convince a local farmer or rancher that West River is not in the grip of drought when soil moisture and rainfall are inadequate during the growing season to support beneficial pasture and crop growth to maturity. In a sense, timing dictates agricultural drought. If the timeliness of the incoming moisture and, to a lesser extent, evaporation, stay out of sync with the planting and growing season over a substantial period, farmers and ranchers begin to feel the effects of drought. In turn, an agricultural drought will have a belated adverse impact on everybody's pocketbook.

Even as the farmers and ranchers struggle with the throes of agricultural drought, the rest of us non-agricultural people, boating on our reservoirs, watering our lawns, and enjoying a cold glass of ice water on a hot day, can remain oblivious of the dry conditions. Only if the rains stay absent long beyond the planting and growing season, city dwellers, sports enthusiasts, hydrologists, and recreationists will then begin to feel droughts effect as well.

It's A Matter of Perspective

There will always be erratic patterns of precipitation across the West River landscape, particularly during the summer months. During any one summer, it's not uncommon to see glaring differences of moisture between the National Weather Service (NWS) office in East Rapid City, where the year-to-date precipitation can remain a few inches above the average, and Rapid City Regional Airport, where year-to-date precipitation can find itself more than 1-1/2 inches below normal.

I reasoned that due to the localized nature of summer scattered showers, we amiably accept this contrast of precipitation. These splash-and-dash, now-and-then showers will drench some areas of the West River prairies and entirely bypass other regions. The luck of the meteorological draw dictates what area gets wet and what area stays dry.

Several years back Kevin Cooper, a past vice president of the local Black Hills Chapter of the American Meteorological Society, e-mailed me his opinion on why the rain gauges showed such a discrepancy. He suggested that some year's precipitation totals between east Rapid City and the airport differ because each location uses a different type of rain gauge. We certainly cannot compare apples with oranges, so how can we effectively compare rain totals from two functionally different rain gauges?

Tipping bucket rain gauge (Google Images)

Kevin claimed the two rain gauges measure moisture entirely different and, therefore, their results will be different. The airport measures rainfall totals using a tipping bucket rain gauge, whereas the east Rapid City NWS office uses the tried-and-true cylinder shaped rain gauge with a standard eight-inch orifice. He said that the airport's tipping-bucket method could introduce significant instrument error, causing it to register less precipitation than what fell from the sky. Here's the problem — when the small bucket fills with rainwater, it tips to empty and then quickly rights itself to its original position ready to be filled again.

The numbers of times the bucket tips mechanically calculate rainfall totals. Kevin pointed out that studies of the tipping-bucket method suggested that its technique underestimates rainfall totals, particularly during heavy rainfall periods. The studies concluded that the tipping and righting sequence of the bucket could not keep up with the downpours typical of summertime thunderstorms. So, when the occasional torrential downpour moved over the airport's rain gauge, some of the rain literally splashed away from the tipped bucket, leaving the rain gauge to record less than the actual rainfall amount.

Albeit the tipping-bucket recorded rainfall amounts now include an empirical correction based on rainfall rates, I still agree with Kevin in that some instrument error cannot be ruled out as a possible contributing factor to the discrepancy, particularly during the summer months when the more drenching downpours occur most frequently.

Simply knowing above-normal versus below-normal precipitation relationships based on a couple of rain gages spaced a few miles apart is not the whole story. One's own perspective on our individual water needs come into play as well. Ranchers, farmers, homeowners, water districts, municipalities, and sportsmen all will offer differing opinions on the year's moisture ranging from above-normal to below-normal and everywhere in between. The uncaring and unemotional rain gauge only records the number of drops that find their way into either its bucket or cylinder. We, the users of that information, decide its relevance to our lives.

Football and the Art of Weather Forecasting

The opening of the football season offers me an opportunity to explain why many weather forecasters use probabilities when forecasting precipitation. As an example, a forecast may include a 60-percent chance of rain for a local football rivalry game. Why use 60-percent probability for rainfall, and not the much more direct, "Yes, it will rain" or, "No, it will not rain?"

First, let's start with the universal truth about the weather: It's not an exact science. It remains impossible to predict how the weather will change for each second of the day and each neighborhood in the city or football field. In part, because the weather repeatedly offers up a complex and mysterious set of continually changing atmospheric marvels across the skies. It is not possible to measure all the minute-to-minute moods and swings of weather with precise accuracy. Finally, even the most sophisticated supercomputers cannot successfully handle and process the billions upon billions of bits of data needed to formulate an exact weather forecast. Some degree of error will inevitably find its way into all weather predictions.

Now enters probability or the field of mathematics that accounts for error lurking along the edges of the precise weather forecast. Probability sorts through the quagmire of weather's imprecision and provides the forecast-user with added information regarding the historical perspective of forecast accuracy.

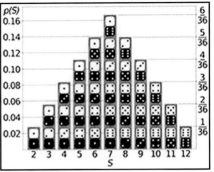

Probability pattern of forecasting a die number (Yahoo Images)

Probability relies upon gleaning some measure of truth from a pattern of a repeated sequence of events. After countless repeated throws of the die, the probability of throwing a seven in a craps game is 6 out of 36 throws of the die. The risk is great that you will not throw a seven.

In our imprecise world of weather, this emerging pattern sets boundaries on the precision of the forecast. By knowing the limits of these bounds, we can assess the risk associated with believing the forecaster. The simple yes/no weather prediction fails to set boundaries, and therefore, fails to evaluate the risk during the decision-making process.

Enough of this heavy-duty mathematical theory dialogue. Let's discuss the probability in relation to football. Picture a fictitious football game, say the Denver Broncos against the Chicago Bears. Just suppose that late in the third quarter during a close game, the Broncos are on the Bears 10-yard line. Based on many years of running the quarterback option, where the quarterback either passes or carries the ball for the score, the Denver coach knows that the quarterback option has an 80 percent probability of scoring a touchdown against Chicago's goal-line defense.

Knowing the probability, or in other words, the boundaries on that play's success rate (i.e., a touchdown will occur 8 out of 10 plays), the Denver coach can now better assess the risk of failure (2 out of 10 plays) associated with running the play. A similar thinking goes into forecasting rain with probabilities.

In the weather game, if the forecast-user knows the probability of rain, they can better assess the risk of getting wet during that outdoor wedding, for example. Risk, coupled with cost, may change some wedding plans. The yes/no forecast does not convey the forecaster's confidence behind the prediction. A 40 percent chance of rain may not alter any plans for a golf game, but it might change the arrangement of a large, expensive outdoor reception.

Some football plays, like some precipitation forecasts, are a sure bet. To illustrate, even if all the Xs blocked all the Os on that quarterback option, with Bob Riggio as the quarterback, the outcome would be 100 percent certain — no score. A sure bet.

When to Cancel Outdoor Activities

Despite powerful computer weather models ingesting loads of weather information from sophisticated satellites and upper air balloon-borne instruments, the yes/no catch-all weather forecast exists only in the minds of the unapprised. Granted that the mathematical inner workings of these authoritative computer weather models include set-in-concrete universal laws, but they also ingest imprecise initial weather observations and not so rock-solid speculative judgments.

Speculative judgments manifest as nothing more than empirical numbers representing educated opinions of why the weather changes based on years of observations, either in the real world or in a laboratory. It is here, within the realm of speculative judgments, that weather forecast model error can too quickly lead to erroneous forecasts. As discussed previously, probability forecasts, such as a 40 percent chance of rain, account for these errors.

Probability forecasts have different meanings to different people. Some interpret the 40 percent rain forecast to mean that 40 percent of the forecast area will get wet, or that it will rain 40 percent of the time during the forecast period. My thoughts say that a 40 percent probability forecast means that 40 days out of 100, your backyard will get wet during the forecast period.

The National Weather Service gives weather forecast probabilities in ten percent increments ranging from 0 to 100 percent. These probabilities culminate from studying weather model output, current weather observations, including wind and moisture patterns, the effect of terrain, like the Black Hills or the Rocky Mountains, long-term weather statistics and the type of precipitation (scattered showers or wide-spread rain).

Since precipitation forecasts use probabilities, let's give the numbers some useful meaning. Zero percent means no precipitation, no way, not today regardless of those ominous-looking clouds on the horizon. Ten percent implies a small likelihood of rain or snow. Twenty percent tells you to expect no precipitation, but not out of the question. Thirty percent says go ahead with outdoor activities but glance skyward on occasion. Forty percent provides you with the comfort of knowing that you will not look stupid carrying an umbrella. When the forecast calls for a 50 percent chance, develop an alternate plan for outdoor activities. Sixty percent, the odds are favorable that Mother Nature might give you more help than the neighboring dog with lawn watering. If a 70 percent precipitation forecast, strongly consider moving outdoor activities indoors, because the chance for no precipitation is only 3 in 10! Eighty percent tells you the wet stuff, or white cold stuff most likely will fall from the darkening skies. A 90 percent forecast means precipitation most certainly will tumble out of now dark skies. One hundred percent means a missed forecast, and I better find another job.

No matter the probability, if it is between zero and 100, the person who did not encounter precipitation during the period would be tempted to say the forecast probability should have been zero percent. On the other hand, the individual who did see rain/snow during the same period would say the chance should have been 100 percent. So, unfortunately, probability says no matter the forecast percentage, it's not fool-proof; the forecast most likely will be wrong to some and correct for others.

Weather Forecasts at Bat

If my forecast called for partly cloudy skies, mild temperatures and light winds, all of which occurred, but the forecast did not mention flurries that fell from the heavens high in the Black Hills, would I be forgiven? After all, three-out-of-four isn't bad. If baseball players, a stock market speculator, or for that matter, a gambler, hits three-out-of-four, they're not only successful; they're wealthy.

So, when my weather forecast — which must contend with the constant vagaries of myriad phenomena – only misses 1 out of 4, am I defamed by the public? Possibly by those folks living in the Black Hills with the unexpected snowflakes falling on their heads. In some place and time, across the vast expanse of the Black Hills area, my forecast attained the 4 out of 4 perfect success rate – possibly at the Rapid City Regional Airport, downtown Hill City or some other place in-between.

The people of Western South Dakota do not live in one tiny place, as a group. They find themselves spread across the Black Hills and the surrounding prairies. Therefore, a local forecast must account for a large geographical area, with changing topography over a relatively extended period. And in the Black Hills or any mountainous region around the country, high and low temperatures can vary as much as 20 degrees or more within a few miles. Rain can occur on the windward side of the Hills and not on the leeward side. Fog can develop over one side of a city and not over the other side. So, like a stopped watch tells the time accurately twice a day, even an errant weather forecast will be on the money somewhere at some moment.

A Weather Forecast Cannot be Right Everywhere

Most weathercasts announce current weather conditions, either at the airport, at the National Weather Service (NWS) office or at some other remote location. Typically, current weather conditions include temperature, dew point temperature, relative humidity, wind, speed and direction, barometric pressure, precipitation amount, and sky cover.

This I can tell you; those numbers on television do not represent the weather conditions outside your home and probably not even in your neighborhood.

It's genuinely ridiculous for a weathercaster to say he can now forecast the weather in your neighborhood. Not possible, because I believe the last time you checked your backyard or walked through your neighborhood you found no weather measuring instruments used in developing the forecast. And, the weather can and most certainly does change dramatically within very short distances and periods, from neighborhood to neighborhood and certainly from backyard to backyard.

Ventilated weather measuring station (Yahoo Images)

The official NWS hourly weather conditions represent one reading snatched during one instant in time from a location about six feet above the ground in a ventilated shelter usually at an airport. That nearly instantaneous measurement may be 32 degrees, while the thermometer outside your window may read 28 degrees and your neighbor's thermometer down the street may read 25 degrees.

It would be foolish to think these three, very different temperature readings should agree with the official NWS observation or any observation even a short distance away because of the weather phenomena called "micro-climates." Thousands upon thousands of micro-climates make up the "broad brush" climate of the area where you live. Micro-climates can be found not only across the town where you live but on smaller scales across neighborhoods, sides of houses and trees, and even different sides of a leaf.

Changes in elevation, vegetation, and surface color that make up your area's mosaic create a nearly endless number of different micro-climates of temperature and moisture. Dark or dry surfaces absorb heat from the sun more efficiently than bright and/or damp surfaces; therefore, one micro-climate appears near the darker rocks and plants where the snow first melts away by absorbing more substantial amounts of solar energy than the brighter surrounding snow.

The angle at which the sun's energy reaches the Earth's surface offers up a noteworthy outcome of micro-climate along the slopes of the hills and foothills. South-facing slopes will typically get more solar radiation than north-facing slopes, particularly in the Northern Plains. Homes located along the south-facing slopes will have a much different outside temperature than a neighboring home situated on a north-facing slope.

Because cold air weighs more (denser) than warmer air, it more easily flows downhill where it collects in low areas, basins, and valleys. Consequently, micro-climates of vastly different low temperatures also can be encountered among two nearby homes that sit on lots with varying elevations.

Vegetation also can create micro-climates by casting shadows and giving off moisture, making the surrounding air cooler and more humid. Urban areas are typically warmer than the surrounding landscape due to the lack of vegetation and vast quantities of energy-absorbing and heat-radiating concrete and asphalt. Heat build-up in urban areas is known as the "heat island" effect.

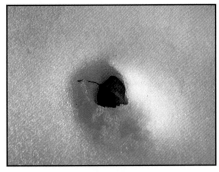

Micro-climate of a dark leaf's warmer temperature melting the surrounding snow

No one who observes or forecasts weather conditions can ever accurately replicate the actual weather across the expansive landscape where you might live. If my weather observation or forecast can convey the proper degree of coldness, hotness, wetness, dryness, or a sense of air movement, I believe I did my job. So, when the forecast calls for 32 degrees, and it turns out to be 25 degrees at your home, don't blame the forecast. Blame those little micro-climates that turn up all over the place.

The Ups and Downs of Weather Forecasting

I know the ups and downs of forecasting the weather, particularly here in the Black Hills. And, may I say, one "up forecast" comes to mind.

I will explain it. My wife and I planned to attend the "Festival in the Park" in Spearfish, SD depending on the accuracy of my weekend weather forecast, which called for an occasional afternoon and early evening shower. Who cares about a few passing showers when you can enjoy such delicacies as pork chop on a stick, roasted sweet corn, and cream puffs with strawberries?

You can imagine my distress when Saturday morning arrived with unexpected damp, foggy, cloudy, and cool weather draped over Rapid City, nothing like I so publicly predicted that Friday night on television. In hindsight, I see that the low pressure system I thought would move through the area did not, resulting in upslope winds, that led to a London-fog-like cloud settling over Rapid City, SD.

My wife, the toughest of all my many critics, informed me bluntly that I busted my forecast, labeling it a "down forecast." "Au contraire," I tactfully argued back. I predicted ideal weather in the town of Spearfish located higher up in the Northern Black Hills. Though the weather in Rapid City did not look like what I expected, our weekend plans must remain focused on the weather in Spearfish. Therefore, with favorable festival weather in Spearfish, my forecast would be an "up forecast" by default, I argued with as much bravado as I dared.

But, each time I dared, the critic responded with such jibes as, "Try looking out a window, Mr. Weatherman." My reputation within the family — not to mention a possible overnight stay in the doghouse — all rested solely on the look of things in Spearfish.

Onward and upward we traveled through the fog with me gazing skyward through the occasional swipe of the windshield wipers for any sign of clearing. You can imagine how relieved I was to find that, driving to Spearfish, our car rather abruptly broke through the clouds into the clear blue sky, much as an airplane might do when ascending upward through the gray and cloudy ceiling.

An upslope cloud layer over Bear Butte, SD (South Dakota Department of Tourism)

I reasoned all along that even though the upslope cloud enveloped Rapid City, it spared the higher elevated town of Spearfish. By the time we reached the festival, the drizzle had vanished, and the sun returned with its warming beams of light. The many festivalgoers welcomed a brilliant day to enjoy the sights, sounds, and smells of the "Festival in the Park."

The dynamics of weather changes happen not only horizontally, as we are accustomed to seeing on East-West-North-South maps, but vertically, as well. And, the changes that occur horizontally cannot hold a candle to the volatility of change that occurs up and down.

As I tried to explain to my wife, with the Black Hills at our doorstep, we must always consider the up-down element when understanding forecast accuracy. How many times have you fought the blizzard in Rapid City when there was not even a light snowfall at a much higher elevation like Deadwood?

Changes in air pressure from low to high or from high to low are the culprit. As the air pressure change, so does the weather fluctuate. Air pressure will also change many times more quickly over a shorter distance going up or down than moving sideways. By traveling westward from Rapid City and upward to higher elevations, we literally drove from one air pressure system, with cloudy, damp weather, into a different air pressure system, with clear, dry weather, clearly proving my point about the "up forecast."

The return trip featured a pleasant drive under mostly sunny skies through Spearfish Canyon. Then, descending from Deadwood toward I-90, again the upslope cloud, fog and drizzle enveloped us. Alas, despite all that scientific explanation, my wife insisted I busted my forecast. What does she know? Now, where can I put all of these arts and crafts purchased at the festival?

Let the Oceans Speak

The last few Super Bowl football games were just that — super, bowl games (except for the 2019 Super Bowl game). Despite the past games' greatness, what I recall most was those never-before-seen commercials.

I could not help but notice that many of those high-dollar advertisements a few years back were "dot com" commercials. Recall the monkey dancing on the metal tub with two guys on either side of it rocking to conga music? After about 10 seconds of this nonsense, the message across the screen read, "We've just wasted two million dollars. What did you do with your money today?"

If the commercial enticed tens of thousands of viewers to go online to play the stock market through the company's web page, then the company did not waste the two million dollars. The sheer numbers of these Super Bowl "dot com" advertisements speak volumes about the immediacy surrounding our lives.

The monkey commercial demonstrates clearly that two million dollars can now be used up in 15 seconds. Secondly, internet technology allows us to go online to gain or lose lots and lots of bucks, or acquire lots and lots of stuff, in a matter of seconds. Like it or not, our current way of life revolves around doing it faster over the shortest amount of time. Buy and sell a stock, or a house, at the touch of the keyboard instead of working through a broker. Instant gratification from playing computer games instead of the slower-pace board games. Eggos for breakfast instead of waffles from scratch.

It's all about instant short-term achievement replacing slow and steady accomplishments. The same hurry-up mentality holds when forecasting weather. In part, due to the invention of lightning-fast supercomputers to calculate complicated equations, and the public's increasing need to know tomorrow's weather at the touch of a button, our ability to forecast the weather in a shorter timespan and more accurately has accelerated significantly since the 1950s.

The resources behind short-term forecasting include meteorologists, applied mathematicians, engineers, and computer programmers and technicians. At no small cost to the public, this global community of atmospheric workers, now interconnect among themselves by a vast network of observational stations and super-fast computers for the primary purpose of providing up to seven days of forecasts.

With that said, it's time to channel more of this vast resource toward improving the long-term forecasts. If farmers and ranchers across the Midwest knew, with some degree of certainty, whether to expect a wet or dry spring, their profits would improve. If the utility companies knew whether the winter temperatures would be below or above average, they too would realize better profits, and hopefully lower costs to the consumer. Municipalities could better plan water usage knowing in advance the summer's expected rainfall. Ski resorts also would be grateful to know about long-term snowfall conditions. The list can easily go on and on.

The discovery of the long-term effect of warmer than usual oceans off the coast of Peru (El Niño) on our long-term weather patterns tells me that the answers to these long-standing questions may very well rest with the oceans. In the past, oceanographers' worked mainly for the military focusing on their needs. Now, in the aftermath of the Cold War, oceanographers' work focuses on the development of longer-term forecasts based on ocean surface water temperatures.

Developing reliable, longer-term forecasts, is a worthwhile pursuit both agricultural and hydrological interests can benefit greatly.

El Niño/La Niña Weather or Not

Over the past few years, popular culture taught us much about the weather phenomena known as El Niño and La Niña and their respective effects on the weather across the United States. Research supports the real possibilities of reliable long lead time climate predictions on temperatures and moisture in the U.S., based on sea surface temperatures measured at the Equatorial Pacific. It works like this: Changes in the ocean temperatures impact the atmosphere over those vast bodies of water, changing climate patterns, locally, across North America and around the globe.

A warmer-than-normal ocean temperature brings about an El Niño event. Here's how it works. El Niño is an unusually warm ocean temperature off the west coast of South America, often reaching its peak warmth about Christmas, thus the name El Niño or "the first child." These warmer-than-normal ocean

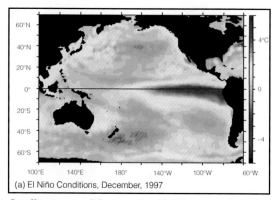

Satellite image of the warming Pacific waters during an El Niño (Google Images)

temperatures change the make-up of the upper-level atmosphere over the water, which in turn disturbs the polar jet stream, that controls the movement of storms across North America. This natural cause and effect progression between the ocean waters, the atmosphere, the jet stream, and storm path, convincingly impact our long-term climatic moisture and temperature patterns around the Earth.

For instance, the weather in Indonesia remains exceptionally dry throughout an El Niño episode, and on the other side of the ledger, warm and rainy conditions dominate the weather pattern in the Southern United States. El Niño episodes may also reduce hurricane activity in the South Atlantic and the Gulf of Mexico by reversing the usual upper-level wind pattern.

An El Niño episode can appreciably benefit us living here in South Dakota during long periods of below average precipitation. You see, historically during an El Niño episode, South Dakota typically gets warmer than the typical fall and winter temperatures and, now the good news, higher than average winter moisture.

Typical polar jet stream pattern during an El Niño event

By comparing nine previous El Niño events, 1914-15, 1918-1919, 1940-1941, 1957-1958, 1965-1966, 1972-1973, 1982-1983, 1986-1987 and 1991-1992, more winter precipitation (rain/snow) occurred during the El Niño events than during the non-El Niño events. On average, Southeastern South Dakota enjoys 147 percent more precipitation during El Niño periods than non-El Niño periods.

Closer to home, Southwestern South Dakota usually receives about 122 percent more precipitation, while Northwestern South Dakota gets typically a smaller increase in precipitation — nearly 104 percent. Unfortunately, El Niño appears to have the opposite effect on the Black Hills, driving precipitation amounts into the below average categories. I believe this result may be due to the Black Hill's topography changing the low-level wind directions, which in turn negatively impacts those moisture-producing upslope winds.

Of interest, the drought of the '30s and '50s ended with the onset of an El Niño period. So, El Niño remains a light at the end of a too long, and too dry tunnel when drought grips the mid-section of the country.

The different, but related occurrence, La Niña, which often, but not always, follows an El Niño event, is characterized by cooler than average sea-surface temperatures in the Central and Eastern tropical Pacific Ocean.

La Niña conditions generally recur every three to five years and can persist for as long as two years. For North America, the most prominent influences

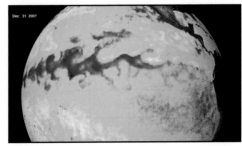

Satellite map of the cooling Pacific waters during a La Niña event (Google Images)

of La Niña happen during the colder months of the year. La Niña impacts the movement and location of our mid-latitude polar jet stream, which during the colder months has greater control of our weather; and both El Niño and La Niña generally peak during the colder months from mid-autumn through spring.

Some believe that the 1998 winter's almost daily rain and snow records across the Pacific Northwest represent the wrath of La Niña. That winter, it rained in Seattle, WA 90 out of 120 days. Mount Baker, WA, received 296 inches of snow or 177 percent of normal. Mount Hood, OR, was buried under 199 inches of snow or 179 percent of normal. Three weeks of official winter remained at the time of those statistics.

In summary: El Niño results in warmer and drier winters in the Black Hills, but wetter conditions for Southwestern and Northwestern South Dakota, while La Niña marks a cooler and wetter winter. Expect wetter conditions during the April through October period during an El Niño and a drier than normal summer during a La Niña.

Weather Futures? Bet on El Niño

Every day we take on some degree of risk in anticipation of a reward. I'll use golf to illustrate. Think about spending an afternoon on an open grassy field in relatively close quarters with up to 150 other club carrying people. Each person uses a club — oh let's say fifty times — to strike a small, solid spherical object into the air.

After doing the math, it's clear that unless all these people have the skills of Tiger Woods, occasionally this swift-moving projectile may drift unintentionally off course toward a group of fellow golfers on an adjoining fairway or tee box. My knowledge of the game tells me to accept a modest risk associated with an errant golf ball when contrasted to the greater reward of spending an afternoon on a beautiful tree-lined golf course with friends.

When taking on new ventures, whether involving possible bodily harm or financial loss, know the risk. Like some of us, I invest a little of my earnings in the stock market. Until recently, my limited knowledge of Wall Street affairs has kept me away from the higher risk ventures that promise better gains, such as commodity futures, though my meteorological curiosity peaked after reading about a new temperature-related weather future on the commodities market.

Weather futures involve temperature-sensitive businesses, who manage their risk from extreme movement in temperatures. Utilities, for example, risk millions of dollars on the whims of nature's monthly temperature swings.

By assuming some of the utility's risk, weather futures offer a prudent investor the opportunity to make a sizable monetary gain, or ... well, you know the flip side. To participate in weather futures, you enter into a contract to either buy or sell the temperature-related commodity at a future date. Here's the risk you take: The anticipated monetary reward rests on your skill to forecast long-term temperature changes at least one month in advance.

Before the 1997-1998 El Niño event, I would have said, "predicting monthly temperature changes with any skill borders on a coin flip." Since the 1997-1998 El Niño event, I have relaxed my previous position on long-term temperature forecasts.

It stands to reason that utility companies can realize substantial monetary rewards by anticipating long-term temperature changes using the knowledge derived from El Niño-determined temperature forecasts. Using this same knowledge in the weather futures market may be to the benefit of a skilled investor. I now concede that the skill of long-term temperature forecasting improves during El Niño years. It also follows that during El Niño years, a wise investment in weather futures may result in a notable reward with a smaller chance of a cost, but never-the-less, still a risk.

During the off El Niño years, my old way of thinking that any long-term temperature forecast remains nothing more than speculation takes precedence. So, I will stay with mutual funds and blue-chip stocks, but who knows — when the next El Niño blows in wetter weather to green the landscape, it may green the pocketbook as well.

Weather Forecast Predicts Cost of Power

Often, we read headlines about power outages in many communities due to downed lines from weather-related storms of many sizes and intensities. Costs to the utility companies to send crews into these communities to restore power can be quite significant. Besides severe weather, there are other weather-related energy costs. For instance, sudden unexpected temperature swings impact the utility tab.

An inverse relationship exists between energy cost and temperature extremes. Plainly stated, as temperatures drop, the utility company's expenses increase with added coal or natural gas costs to generate more electrical power to keep us warm.

Typically, the utility company spreads these additional coal or natural gas costs among its customers. It's safe to say that, after reaching into one's wallet to pay last month's winter-generated electric bill, you quickly acquire an indelible impression about that month's wintry conditions.

The amount due reflects not only apparent cold temperature energy costs but also mirrors not-so-obvious weather-related energy costs, such as how well the utility company predicts weather events. You see, the better the utility company can forecast changes to high and low-temperature values, the more money you keep in your wallet.

To keep operating costs down, utility companies work hard to improve operational efficiency through better temperature forecasting. Having the ability to predict accurately short- and long-term temperature swings allow the utility company to anticipate future hour-by-hour load demands. Knowing just what energy load demands will be ahead of time can save the company and its customers a lot of money.

Electrical power lines

Here's how it works based on a utility company I worked for in Texas: Utility companies base their daily needed energy demand on a load forecast of various weather variables. Keep in mind that it takes many hours and a lot of dollars to bring one power generator online to accommodate the expected power demand.

The load demand is highly sensitive to temperature fluctuations followed by cloud amounts, precipitation, and winds. A five-degree error in the load temperature forecast can result in a costly start-up generation of unneeded electric power.

For instance, if an inaccurate load-temperature forecast calls for starting up three power generation units when only two are needed to meet demands, the utility company must either sell the additional power to another utility or pass the additional start-up cost to their customers. To give you an idea of the costs involved, a typical outlay to fire up one coal-fired unit can run in the tens or hundreds of thousands of dollars a day.

On the flip side, if an inaccurate load-temperature forecast calls for starting up only two power generation stations when three were necessary to handle the power demands, the power company would have to buy that additional power from another company.

It does not take a mathematical wizard to understand the high value of an accurate temperature forecast to startup-shutdown of power generation units. A conservative annual estimate of weather-related error costs due to start-up-shutdown of generation units runs into the tens of millions of dollars. Succinctly, a large percentage of those dollars can be saved through slight improvements in short-term forecasts.

So, a good forecast goes a long way in saving the power company costs, and ultimately, keeps a little more money in our pocketbook.

Uncertainty Part of Flood Forecasting

In April 1997, the Red River of the North spilled over its banks, causing the worst flood in the history of North Dakota. Damages to the communities along the river totaled $1-2 billion, with the more significant loss occurring in Grand Forks and East Grand Forks.

The 1997 flood in Grand Forks, ND (Google Images)

The Red River Basin, with its relatively flat landscape making it especially prone to flooding, flows north through North Dakota and Minnesota and into Southern Manitoba, Canada. I understand that it's counterintuitive to believe the river flows to the north, but it does, forming 440 miles of the North Dakota-Minnesota border before flowing into Canada.

Somewhat unique to the Red River, spring flooding frequently occurs due to the northward migration of snowmelt. As the spring-meltwater flows north into the colder portions of the river, it soon becomes obstructed by the remaining river ice located in the colder northern climes. This natural damming causes the Red River to rise over its paltry banks and spread across the region. Snowpack, timing, rate of melting, additional precipitation, and soil moisture, all add to the flood equation.

During the winter of 1996 and spring of 1997, the Red River Basin received record amounts of snowfall and battled through eight blizzards. Severe ice storms caused big-time power outages in the Grand Forks region, placing flood efforts on the back burner for a week or longer, adding one more component to the impending disaster.

The National Weather Service (NWS) predicted a 49-foot crest or a foot and a half above the previous record. More importantly, the forecast called for a crest one foot below dike capacity of 50 feet. In the days that followed, the river crested at an astounding 54.11 feet — 5.31 feet above the highest crest ever recorded.

From April 14 through April 18, 1997, hundreds of volunteers tried in vain to fortify the dikes for levels beyond their 50-foot limits. As the beleaguered workers watched the fortified barriers breach and crack, they knew the river had won. The floodwater spilled over the embankments and into the cities of Grand Forks and East Grand Forks, cutting off those communities from dry land as the bridges dipped below the water levels.

By Sunday, April 20, most residents were evacuated, leaving only empty shells of homes, schools, offices and hospital buildings behind to withstand the rising water of the now-bloated Red River. Amazingly, with water everywhere, Grand Forks' downtown caught fire destroying 11 buildings, some of significant historical value. Not until Wednesday, April 23, did the river begin to recede, leaving the sodden remains of lives and memories scattered across the city.

As the river went down, accusations went up. In the words of the mayor of East Grand Forks, "They [NWS] missed it, and they not only missed it, they blew it big." Sure they "blew it big," but here's one reason why.

History never before recorded or even observed a catastrophic event of this magnitude, adding further inadequacies to the then current river forecast mathematical model. Many equations deeply embedded in a forecast mathematical model depend upon observed and well-documented events from the past. By incorporating past events, the models can produce a better and more rigorous forecast of future events. Unfortunately, new record occurrences that happened after any model-included record events, manifest extrapolation errors beyond the boundaries of the model's capabilities. This extraordinary flood far exceeded any historical floods on record and, amazingly, the model responded with a new record forecast, albeit well short of the actual crest.

For this reason, and others, any mathematical model forecast will always carry with it a certain degree of uncertainty, especially when asked to predict a never-before-observed event. Better understanding and communication of river forecast uncertainty, between the NWS and the local authorities and decision-makers, should be the one valuable lesson learned from the 1997 Red River flood. For that matter, communicating forecast uncertainty should be applied to all NWS forecasts.

NOAA Gets Weather Out Fast

I like to think I'm an easy-going person. After all, I raised three kids who managed to go through adolescence without harming themselves. Lately, though, I've noticed certain events that are chipping away at my perseverance.

Case in point: The drivers in front of me appear to be driving much too slowly for my taste. Checkout lines now seem to move at a snail's pace. And, a meal is not a meal unless I get it from a drive-through window. After some thought, I believe the fault does not lie with my patience; it lies with my perception of time.

Today's fast-paced technology significantly shortens that interval within which we tolerate our surroundings. Consequently, in today's "just add water" world, any disregard of one's time begins to erode our patience.

Once we adapt ourselves to doing things faster, it's challenging to return to the "good old" slower-paced days. For instance, today the museum's exhibit rotary telephones, usually found alongside the typewriter display. Can you imagine dialing a nine-digit telephone number on a rotary phone, when you can now make that call by simple voice command? Why squeeze oranges when you can buy orange juice by the carton? And money instantly appears out of a hole in the wall known as the ATM.

Undoubtedly, speed counts in today's world and, interestingly, our brain can handle whatever we throw its way no matter how fast. I believe our brains have a nearly unlimited capacity to assimilate information quickly, and weather information is no exception to this "Riggioism."

The public cannot get enough information about weather. In nearly every source of news, whether newspaper, radio, television, Internet, smartphone, web sites, or the corner café, the weather remains the topic of paramount interest.

And why not? It impacts our livelihood, well-being, state of mind, and health. Our fascination with weather will always last because we find ourselves stepping outside into the weather daily, or worse, severe weather can step inside our space.

Aside from the typical news media and the internet, all urbanites, and an increasing number of rural residents, can obtain first-hand the forecast, weather advisories and much of the data collected by the National Weather Service almost instantly through the government-operated radio network known as the National Oceanic Atmospheric Administration (NOAA) weather radio broadcasts and government web sites.

In past years, there were more than 480 NOAA Weather Radio broadcast stations in the 50 states. Surely there are currently many more. NOAA Weather Radio disseminates current weather conditions, the state of the atmosphere, and local forecasts. More importantly, it broadcasts watches when conditions are favorable for severe weather, and warnings when severe weather is occurring or imminent. Weather observations, both surface and aloft, can be found easily on government web sites.

To appease our desire to know 'immediately' what's going on, the Rapid City NOAA Weather Radio uses a program that turns written words into a synthesized-voice broadcast. The computer's voice may not sound as good as your favorite weather anchor, but it significantly reduces the time it takes to get severe weather watch and warning information to the public.

And when it comes to severe weather, at once is never fast enough.

Weather Service Not Liable for Forecasts

If McDonald's can be sued for a hot cup of coffee, then surely the National Weather Service (NWS) can be sued for a wrong forecast. Before dialing up your lawyer, understand that a successful lawsuit against the NWS for an inaccurate or inadequate weather forecast is highly unlikely.

The U.S. government could care less about lawsuits — no questions asked. Something called "sovereign immunity" prevents the government from being sued because "the King can do wrong." Because the King and government can do wrong, Congress passed the Federal Tort Claims Act (FTCA), making the government liable for loss or injury.

Even with the passage of the FTCA, prosecuting the NWS for a wrong forecast can be most challenging at best. Here's the catch. The NWS can be exempt from any obligation or penalty under the FTCA based on the exercise of two well-meaning clauses – the "discretionary clause" and the "misrepresentation clause." Applying either of these two clauses to a lawsuit against the NWS, dismisses the lawsuit.

The "discretionary clause" shelters most, if not all, weather service activities. You see, NWS forecasters use discretion in the function of their day-to-day job to create a forecast. Forecasters need to decide what information to use to forecast the weather. Also, they gather and select weather observations and other data from various sources. The forecaster chooses which computer model best reflects the current weather situation, and then they must decide how best to disseminate the forecast to the public.

All of these everyday activities require the exercise of judgment and decision-making by the forecaster; therefore, excuses them from any and all adverse consequence resulting from a forecast derived from this discretionary process. The "discretionary clause" keeps the NWS on a steady platform by preventing frivolous lawsuits from changing staffing levels, budget resources, or technological needs within the many weather-service offices scattered across the country.

Without some degree of immunity in place, lawsuits against the NWS would almost certainly water down forecasts to the point where they would be of no use to the community they serve. Forecasts might read something like, "It may or may not snow today." It's likely that without these clauses in place, demands by successful lawsuits would ultimately bring the NWS to its knees.

Because weather forecasting is an inexact and indeterminate science that varies significantly from street corner to street corner or from pasture to pasture, most users of the NWS forecast almost always have a beef about its accuracy or completeness. Keep in mind that the complaints probably are best discussed in the coffee shop and not in the courtroom.

Dan Lutz

The somewhat steady tilt of the Earth, coupled with its consistent speed around the sun and spin on its axis, methodically delivers the change from summer-like days to fall-like days to winter-like days and spring-like days, better known as the changing seasons.

🍃 THE SEASONS

Once a year, the climate across the Northern Plains officially changes from summer to fall to winter to spring and back to summer again. These seasonal shifts of our Northern Plain's climate happen on cue, according to nature's astronomical clock.

On the day the summer season officially ends in the Northern Hemisphere, the sun crosses directly over the equator on its southward journey high in the sky, formally starting the six-month period of the year when nighttime hours exceed the daytime hours. Days with warm nights, hot afternoons and scattered thunderstorms will progressively transform into days with much cooler nights, mild to cold afternoons and, on occasion, an overcast sky dripping with steady rain or snow.

Six months later, the sun will again cross the equator, but this time it'll be moving northward during the spring equinox or the counterpoint to the autumnal equinox. And in our part of the world, i.e., the Northern Hemisphere, the warm nights and hot afternoons return slowly, melting away the snow-covered ground and kicking off those afternoon showers and widespread rains so necessary to once again green-up the Black Hills and prairies, and replenish the creeks, lakes, and reservoirs.

The somewhat steady tilt of the Earth, coupled with its consistent speed around the sun and spin on its axis, methodically delivers the change from summer-like days to fall-like days to winter-like days and spring-like days, better known as the changing seasons.

I like to believe that these steadfast cyclic rhythms of nature compel us to take notice of our endlessly changing surroundings and for us to take some time to reflect on and accept the transformation taking place. For example, you surely would find yourself thinking about how the new spring days will be different from yesterday's winter days and possibly different from the upcoming summer days. We can count on nature to once again bring us the fall colors, the winter snows, the spring flowers and summer's warmth, and maybe the changing seasons remind us of something more essential to our lives.

Like the seasons, life changes are inevitable: loved ones pass; job status change; health issues may arise. The difficulties and mistakes of the past cannot be reclaimed and remain untouchable for a redo. Perhaps, with a little knowledge gleaned from the past, coupled with a bit of forethought and determination, we too, like the changing seasons, can count on more certain days ahead.

Fall Arrives at Once

The autumnal equinox begins the seasonal transition when summer officially ends, and fall begins. On the calendar, the change usually occurs on September 21 — give or take 24 hours. During the fall season, the spring and summer's energy to grow and nourish life, whether animal or vegetable, fades and returns to the Earth. The vibrant green leaves of the oaks, aspen, cottonwood, and elm turn color, die and fall to the Earth. Many animals return to the Earth as they bed down underground for the winter. I believe it's safe to say that all living creatures lose some energy to the Earth and begin to slow down, as the landscape changes and our metabolism slows.

It's interesting to note that many cultures consider this time of year — the autumn season — to be more of a time of repose, relaxation, and realignment, as the nights remain longer than the days. Thoughts not only turn toward our surroundings but inward as well, with a focus on our change from yesterday to today and into tomorrow.

The transition from one season to the next does not ripple across the globe. It happens all at once. The seasons take their cue from the heavens and not some fixed Earth-bound clock. These annual transitions of the four seasons occur on cue according to an astronomical timepiece.

When the seasons' transition from winter to spring and from summer to autumn, also known as the equinox, both the Northern Hemisphere and the Southern Hemisphere enjoy approximately 12-hour-long periods of sunlight, followed by 12 hours of darkness. Thus, the word "equinox" means the equal length of time for daylight and night darkness.

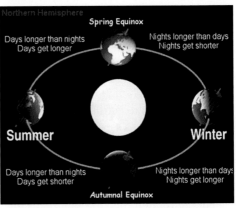

The seasons' transition cycle (Google Images)

On these equinox days, the sun will set nearly due west and rise virtually due east. And, over the following few weeks, the sun's location on the horizon as it sets and rises will show its most significant change from one day to the next day. A careful observer here in the Black Hills after the autumnal equinox will notice the sun rising and setting farther and farther south of due west. Also, after the autumnal equinox, our daylight will transition from a more extended period to a shorter period than our neighboring states to the south.

Numbers Say Fall Warmth is Normal

During autumn 1998, the October and November day-to-day temperature extremes changed like the steep climbs and speedy drops of an amusement park roller coaster ride. Afternoon high temperatures ranged from 78 degrees on October 8 to 29 degrees on November 4. Morning lows ranged from 41 degrees on October 9 and October 14 to just 13 degrees on November 11.

Most of us living here in the Black Hills in 1998 would agree that that autumn appeared unseasonably mild. We will remember the warm Thanksgiving holiday with mercury readings in the lower 60-degree Fahrenheit range. And we won't forget the record 70-degree Fahrenheit high temperature recorded only eleven days before Christmas. Just as a squirrel stores away food during the autumn to help survive the winter, we lock away the short-term memory of these warm autumn days for immediate recall on an as-needed basis during the undeniable cold of the winter to come.

Surprisingly, though, after applying some basic statistics to the hard numbers I collected and logged on a spreadsheet, what becomes apparent may cause us to question our memories of a warm mild autumn. That year's average high temperatures for October and November were not above normal; In fact, the average high temperatures were slightly below normal, while the average low temperatures were only a few degrees above normal.

Autumn temperature extremes in the Black Hills can rival the most exciting roller coaster ride you would like to remember. My all-time favorite roller coaster is the Great White Shark at Texas Sea World, with all the ups, downs, twists and turns a grown man can handle within the 30-seconds from start to its merciful finish.

But after averaging the peaks and valleys, our roller coaster becomes more like the kiddy train ride that used to chug around Storybook Island, with its smooth turns on flat rails. Average smoothing declared the autumn of 1998 as near normal.

It's a simple fact of life that years from now, all the extraordinary temperature peaks and valleys we experience will lose themselves to statistical mediocrity. But for the short term, despite the statistical averaging, let's not forget those sunny skies, long shadows and many unseasonably warm days and nights across Western South Dakota.

Winter's solstice marks the beginning of the country's coldest season. Symbolically, from our point-of-view from here on Earth, the sun has completed its journey southward in the sky and stops to rest on this day before beginning the trip back northward, toward its highest point in the heavens.

When the winter season or any South Dakota season arrives, expect the temperature roller coaster to leave the platform like those 1998 autumn temperatures. So, let's get on, buckle up, and enjoy the temperature peaks as well as the valleys. And let's not forget those past delightful and stimulating past seasons with those strewn-in unseasonably warm and cold days to breakup the wearisome uniformity of the season, despite what future statistics will portray as just average.

Why Autumn Leaves?

All scientists, including the atmospheric scientists, rarely if ever use the phrase, "just because." Scientist's wiring dictates that their need-to-know moves beyond nature's face value and into the minuscule workings of why things happen. By exposing all of nature's mysteries, we lose some of the joyful spontaneity and unguided imagination triggered by our surroundings.

For instance, when a child asks: "Why a rainbow?" What better way to stifle the child's imagination than by describing how spherical globules of water refract white light separating a spectrum of colors and bending them across the sky. I like to think an answer like "just because" from time to time allows the child's fantasy to take flight.

I cannot think of a better way to spend an afternoon than walking, riding or driving through the Black Hills soaking up the beauty of the season's autumn colors. Nature's own tapestry peaks with the greens, golds, oranges, and reds backdropped by a canopy of a deep blue sky. Assuredly, while in this moment surrounded by nature, "just because" should be the much more appropriate answer to "why" than some dissertation about biochemical reactions.

Fall colors in Jackson Park, Rapid City, SD

But for those of you who have already experienced the autumn season's splendor in ignorant bliss, I will now go beyond the "just because" and into autumn chemistry. However, words of warning to those who have yet strolled through the park or experienced Spearfish or Vanocker canyons during the autumn season—I will only explain the "why" behind autumn leaves turning colors. You will fully appreciate the "just because" after a short autumn ride (preferably by motorcycle) through one of those canyons.

Every autumn across the Black Hills, diminishing daylight hours coupled with lowering temperatures begin the spectacular color show. The leaves stop using numerous cells containing chlorophyll, which gives the leaf its green color, to make food. The chlorophyll breaks down, the green color disappears, and the yellow to orange colors become visible and give the leaves a part of their fall brilliance.

At the same time, other chemical changes that form additional colors through the development of red anthocyanin pigments or the yellow carotene pigments may occur. Some mixtures give rise to the reddish and purplish fall colors of trees such as dogwoods and sumacs. The autumn foliage of other trees like the sugar maple shows brilliant orange, while birch shows bright yellow. Others, like many oaks, display mostly browns.

Temperature, light, and water supply influence the intensity and duration of autumn's color display. Cold temperatures that stay above freezing will favor brilliant reds in maples, while an early frost will notably diminish the bright red color.

A wet growing season followed by a sunny autumn tend to increase the intensity of fall colors. The best time to view the autumn colors would be on a clear, dry, and cool but not freezing day. Strong winds capable of stripping leaves from the trees and bushes will also shorten the viewing time of nature's color display.

So, there you have it: Jack Frost had little to do with this autumn's color exhibition. The colors are due to the mix of varying amounts of stuff like chlorophyll residue and other pigments in the leaf triggered by the sun's lowering angle during the autumn season.

Now, with that said, purge this column from the gray matter and enjoy the gold, red, orange and yellow matter found throughout the Black Hills area each autumn season.

Ah Autumn, Furniture Sales

Well into the autumn season, you cannot help but notice and, hopefully, appreciate the sweeping changes going on across West River. You see the leaves turning colors and tumble to the ground, feel the chill of the northwest wind, watch high school football on Friday evenings, and — oh yes — marvel at the apparent prosperity of furniture stores. That's correct; I'm convinced home furnishing sales peak at this time of year, right along with the gold, red and yellow colors evident throughout the beautiful Black Hills' canyons.

If you doubt this apparent autumnal home-furnishings buying spurt, let me prove it by stepping through the traditional scientific method of experimentation. First, I will state the hypothesis or best guess: "People buy more furniture during the fall season than any other time of year." The second step of my "comfy couch" scientific study tests the hypothesis to prove it true or false.

The test took place last weekend during an impromptu shopping excursion with my wife. Although we have a house full of furniture, for some bizarre reason we found ourselves drawn to furniture stores. At each home-furnishings establishment, we encountered crowds of other potential "comfy couch" buyers, all with the same glaze-eyed, confused look on their faces.

After two days of analyzing the data, I reached my conclusion. Based on the scientific method and a limited license to generalize, I accepted the hypothesis. Autumnal characteristics — like shorter days and longer shadows — affects our psyche, triggering a sub-conscience craving for comfy couches.

If you think about it, this now proven hypothesis makes reasonable sense. As the sun retreats to lower and lower heights in the sky, all living creatures, including us humans, sense the coming of winter and its frigid assault. As the days grow shorter, the need to "nest up" and prepare for the cold becomes a psychological obsession.

Nature prepares horses against winter's chill by growing their coats longer and fuller. Squirrels gather nuts to feed themselves through the winter to help stay warm. Throughout the quiet autumn days, even wild birds know enough to fly south to warmer climes. It's all part of nature's plan to nest-up cozy and warm before the winter's cold and snow. As I have shown through my careful scientific approach, we humans follow similar seasonal-based rituals as we wonder through furniture stores looking for that secure-feeling and restful sofa. The comfy couch syndrome buried in many Northern Plains folk's constitutions helps to satisfy this compelling urge to nest.

If you are now one of the many recent buyers of a comfy couch, no need to worry about the now-added clutter in the living room. Nature has a plan for that, too — the garage sale. It goes like this — next spring, all living creatures will desire to "nest down." Bears exit their dens, birds abandon their nests, and humans have garage sales. No need for scientific reasoning here, "nesting-down" has something to do with the notion that less is best during the longer and warmer days of summer.

We have completed another of nature's circles of life. Each autumn, people flock to furniture stores with this mysterious desire to buy cozy furniture. We finish the loop each spring when a perfectly good piece of furniture ends up at the garage sale, restoring comfy couch balance once again. At least until the following fall.

Conditions Gang Up in Spring

It's during the spring when rain clouds evolve into thunderstorms that often progress into severe weather across the prairies and Black Hills of Western South Dakota. Severe weather can occur at almost any time, but May and June are the peak times for booming thunderheads to unleash their large hail, damaging winds, slashing downpours and, on occasion, tornadoes.

Wind shear in non-rain producing cloud over the Black Hills, SD

In the spring, the sun climbs higher, and surface temperatures begin to soar. Morning frost becomes a memory, and my thoughts turn toward vegetable and flower gardens, riding motorcycles, playing golf and fly fishing. Increasing warmth and the seasonal migration of the jet stream will cause me to keep the threat of severe weather in mind, as well.

During the months of spring, the upper-level jet finds itself over the Northern Plains, producing different wind speeds at different heights. This vertical increase in wind speed allows rain clouds to rise to great heights by causing them to tilt or shear, which prevents the falling rain from killing the clouds' updraft.

Also, the mid-level jet comes into play by pushing dry air from the desert Southwest farther north, offering up evaporative cooling high over the Black Hills. With warmer air near the surface, the mid-level cooling causes even greater instability in the growing rain cloud. Finally, the Bermuda High pressure airmass, in the Atlantic Ocean brings abundant Gulf of Mexico moisture northward around its western edge in a monsoon-like surge.

As a rule, when these ingredients for our severe weather recipe come together, powerful thunderstorms form during the late springtime afternoons, when heating exerts its maximum "boiling" effect on the atmosphere. Rarely does the West River region of South Dakota experience the wrath of severe thunderstorms during the mid-morning hours.

For the most part, severe thunderstorms in the Black Hills produce large hail and damaging winds. Less often, some of the more potent storms produce tornadoes. Officially, Western South Dakota thunderstorms most often spawn "weak" tornadoes with wind speeds of less than 112 miles per hour and a short life span on the ground, usually measured in tens of minutes.

Do not let that fact give you a false sense of security. Many Western South Dakota tornadoes classified as weak may, in fact, be "strong" to "severe." Surveys of structural damage currently determine tornado strength. Here in South Dakota, on the open expanse of rolling prairies, few structures end up in the path of a tornado. Consequently, many tornadoes end up classified as weak when, in fact, they may have been extremely powerful twisters.

Springtime rain shower weather in May and June is frequently beneficial for the farmer, rancher, and gardener, and often exciting for the severe-weather buff. So, keep in mind that those same early afternoon rain showers that nourish your garden can lay it flat later that evening under a swath of hail.

Summer's Heat from the Dogs

Here's a news flash — summers in the Black Hills can and do become exceptionally hot. During the summer of 2002, 10 of the June's summer-season days reached or exceeded 90 degrees, with three of those days exceeding 100 degrees. And the July daily summer temperature readings only got hotter. Of July's 31 days, 24 days reached or exceeded 90 degrees, with six of those days reporting temperatures of 100 degrees or higher.

That year, the average Black Hills July temperature reached 77.9 degrees Fahrenheit, or 6.2 degrees Fahrenheit above normal. It's remarkable for any weather station for any time of the year to attain that large of a temperature departure from the average. Looking at the daily extremes, the average high temperature for July topped out at 93.5 degrees Fahrenheit or 7.5 degrees Fahrenheit above normal. Also, the average July low temperature came to 62.3 degrees Fahrenheit compared to the normal of 58 degrees Fahrenheit.

To underscore those extraordinarily hot temperatures recorded during July 2002, the National Weather Service classified it as the warmest ever when compared to the past 100-plus year's average July temperatures. Regardless of this "hottest-ever" distinction, only one day — July 15 — recorded a new record-high temperature. Imagine, if you will, how far above the average high temperature the other days of the month must have been.

If ever there was a summer that represented "the dog days of summer," 2002 must be it. Which begs the question: Where did the phrase "dog days of summer" originate? Well,

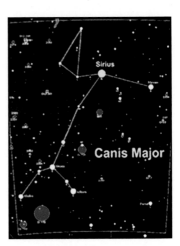

The star Sirius and the constellation Canis Major (Yahoo Images)

it originated with the ancient Egyptian belief that the star Sirius found in the constellation Canis Major, or the major dog, added to the summer's heat. You see, besides the Sun, Sirius is the brightest star in the Northern Hemisphere, and in Egypt, Sirius shines for most of the summer. And because of its brilliance, the Egyptians believed that the sun plus the extra light from the nearby star caused the summer's heat.

The star's name comes from the Greek word meaning "searing" or "scorching." Sirius endures as the most notable star of the constellation Canis Major, which represents Orion's bigger hunting dog, and is therefore commonly referred to as the "dog star." Sirius' mass measures in at more than twice that of the sun. It shines hotter, about 9,400 degrees Fahrenheit, and, consequently brighter, or 25.4 times as luminous as the sun. Its close location to Earth, a mere 8.6 light years away, also makes it appear bright to us.

As if one needs directions to find the brightest star in the sky, from the constellation Orion, the brilliant Sirius can be found by looking down and to the left. The good news is that the only time you will be able to observe the "dog star" is during the cool of the evening.

Weather Theory Well-Grounded

With the passage of the winter solstice, the sun will rise over the eastern horizon progressively earlier and set over the Black Hills progressively later. The sun's track across the sky will appear to be a little higher with each passing winter day, allowing for a more direct and temperate route to the Earth's surface.

It stands to reason that a daily increase of the intense and continuous stream of energy from the sun that enters the envelope of air surrounding Earth would translate to a quick return of warmer weather. But of course, the weather is never straightforward.

Earth's large-scale weather patterns change slowly in response to changes in the solar cycle. Ancient Egyptians, Mayans, Incas, and thousands of years later, the great minds of Copernicus, Galileo and Newton demonstrated that daylight hours march to an exact, precise and predictable beat as the Earth spins on its axis and orbits the sun.

On the other hand, the Earth's usual daily temperatures march to a drummer slightly out of step with the Earth to sun relationship. Despite the day-to-day increase of daylight hours and, subsequently, additional warming sunshine, the daily average high and low temperatures will continue to fall to their annual minimums until early January or weeks after the winter solstice. The record cold for Rapid City occurred on February 11, 1899, 52 days after the winter solstice, which plausibly should mark the lowest temperature point in the season.

Sir Isaac Newton, who explained several of nature's fundamental laws, told us that action set in motion tends to stay in motion unless acted upon by some other outside action. Green Bay Packard coach Sir Vince Lombardi had a similar law: a football team moving down the field to score a touchdown will indeed score the touchdown unless a more powerful outside action, i.e., the opposing team, can stop them (all bets are off if the opposing team is "da Bears").

Snow covered ground in the Black Hills, SD

The same holds for downward sliding average temperatures, for they too, will continue the downward plunge until some action can effectively reverse the trend from colder temperatures to warmer temperatures.

The necessary opposing action to reverse the falling temperatures is not so much the immediate return to longer days, but the ground's ability to quickly absorb and retain enough heat. A cold and frozen ground loses much of its ability to warm rapidly, keeping the average minimum air temperature three to four weeks behind the minimum solar radiation the winter solstice provides. Expect an even slower return to warmer temperatures, if the ground has a deep frost or snow cover.

So, look toward the ground if you want an indication of how long cold winter temperatures will stay around. In a year with below-average snowfall, we can expect an earlier return to warmer temperatures. With piles of snow remaining Earth-bound long after the days become longer, colder temperatures will persist longer.

The groundhog may be smarter than I give him credit for. After all, he more than any other mammal keeps a very close tab on the ground's day-to-day condition.

Waiting on Old Man Winter

Winter comes about by the tilt of the Earth. The Earth spins around on a 23.5-degree tilted axis relative to the sun. Between the autumn equinox (September 21 or thereabouts), and the spring equinox (March 21 or thereabouts) the Northern Hemisphere tilts away from the sun, bringing about shorter days and longer nights.

This downward shift of daytime sunlight lessens the impact of solar heat to warm the northern half of the globe and, conversely, delivers more nightfall to cool off — radiation cooling. During those winter months, the Northern

Hemisphere will always remain slightly tilted away from the sun. It stands to reason that, due to shorter days and longer nights, the Northern Hemisphere loses more heat than it can accumulate and, therefore, becomes colder and colder with each passing day. Not only growing more frigid in the Northern Plains but even colder farther north in the Arctic.

Most years, Arctic-borne cold-air masses slip southward into the Northern Plains around the winter's solstice, giving us a gradual taste of colder and colder winter days. Some years, however, the location of the jet stream holds that initial chunk of winter's cold air firmly in place over the Arctic, allowing the Northern Plains to enjoy an extended period of warmer-than-normal temperatures. Unfortunately, during this unseasonably warm period in the Northern Plains, the Arctic regions become even colder reaching levels well below average. When the jet stream finally moves even slightly, losing its grip on the frigid, dense mass of Arctic cold air, it can break loose and quickly slide southward, chilling to the bone, not even the northern states, but most of the lower southern states east of the continental divide.

Unfortunately for us, the Northern Plains states will take the opening blow. Oh yes, there are years when the throes of Old Man Winter may not arrive across the Black Hills on time, but when the first real taste of winter pours out of Canada, it probably will not be a taste to savor. Often that first polar express will bring wind chills of 20, 30 and even 40 degrees Fahrenheit below zero.

When the first polar express arrived in Rapid City on February 3, 2019, it carried with it bitterly cold temperatures and wind chills far below normal, where they remained far below average for the rest of the month. At the time of this writing, February 2019 turned out to be the third coldest February on record replacing the previous third coldest February in 1891.

Invasion of the Mutant Snowflakes

The start of the first significant and extended snowfall generally waits until late autumn or the early winter season when quiet snowflakes cover pine trees with winter's white fluffy decoration. Often the following snowflake-filled day, after the poetry of those silent snowflakes fade, along with 24 hours' worth of additional soft snowflakes, we find ourselves shoveling sidewalks and driveways, and scraping the now bygone snowflake-loveliness from the windshield.

A heavy snowfall event in Rapid City, SD

By the next day, those same annoying snowflakes, and their newly arrived friends proclaim that to better travel the byways and highways of this great state; I must purchase appropriate tires. By the third day, those diabolical flakes, along with yet another day's worth of snowflake reinforcements, invade my personal space, screaming at me to leave my warm bed and grab the new snow shovel and scraper. Those fiendish flakes wish to meet me outside in the cold to do battle on the walk, the driveway, and the car's windows.

I am not alone in my battle with the lords of the flake, as the cities and counties across the State, and the State itself, muster the maximum muscle of their plows to push the hordes of fallen flakes aside. In winter, the loud collected voices of a countless number of pinhead ice crystals boomed across the Black Hills region, causing tire dealers and hardware stores to live and prosper.

With patience and perseverance, there will always be a time we all can rejoice to say that we have won the battle over the throngs of invading mutant snowflakes that first arrived as welcomed "quiet fluffy white stuff." Our lives will again return to the ordinary. Joggers will jog. Construction workers will construct. Newspapers will be delivered. And, I'll walk my dog Wally in Jackson Park.

So, from where did the snowflake invasion originate? You see, the snow has many forms and shapes that allow it to mutate beyond the quiet, white, fluffy stuff. Sometimes it can be gentle, but it also can be hard and icy. The atmospheric temperature profile and cloud conditions within which the snow formed dictates the form snow takes when it reaches the ground.

Snow grains/pellets (Yahoo Images)

Snow grains, snow pellets or grapple suggest that the snow originated in the cloud as water droplets, becoming super-cooled liquid water high in the cloud well above the freezing level, followed by colliding with and freezing on ice crystals, and finally falling to the ground. Ice crystals indicate the snow perhaps skipped the water phase altogether and went directly from water vapor to ice crystals by forming on dust particles. Even under a cloudless sky, ice crystals can appear out of clear air with cold enough air temperatures.

Snowflakes, the most common type of snow, form in large tall clouds that can accommodate places cold enough for ice crystals to form and other lower and warmer regions to allow for a partial melt of the ice crystals so they can cling together with other ice crystals to form snowflakes.

Snowflakes formed by ice crystals (Yahoo Images)

It is an article of faith that there are no two snowflakes that are alike, but the crystals types do fall within similar classifications. The cloud's temperature and water vapor content dictate the final appearance of the finished snow crystal. The crystal shape can be in the form of plates, needles, feathery dendrites, or columns.

Most importantly, when near-freezing air temperatures create wet and heavy snow — snowball battles, snowmen, and snow forts cannot be far behind. The best snow-packing for snowballs, snowmen, or snow forts comes about when the near-surface temperature stays slightly above the melting temperature or slightly above 32 degrees Fahrenheit.

What If Humans Napped Winter Away?

Oh, this ingenious, yet the gloriously simple question of nature will surely dance through our heads during one of our daily rites of winter, like the boot stomp and boot removal at the front door threshold. Or, kicking off of our vehicles that first big ball of greyish crud that accumulates repetitively behind the tires. Oh yes, we will ask ourselves why nature gave wild animals the perfect winter sanctuary, and not us.

NOVA, National Geographic or the Nature Channel show us that hibernation applies to only a relatively few living creatures. For them, hibernation is an annual natural defense mechanism. If it's too cold to hunt for food and water, then just sleep it off for a few months. Many small- and medium-sized mammals that call South Dakota home enter this prolonged and controlled state of apparently comfortable sleep. Why did not a benevolent Creator grant we mortals the ability to confuse Jack Frost with The Sandman? After all, we stand tall on top of the food chain as the supreme creatures.

Our superior intellect should entitle us to add those extra annual pounds, guilt-free, by gorging ourselves with our favorite foods around Thanksgiving. Oh yes, guilt-free, because the poundage will surely melt away during our long winter nap. The body will slowly and meticulously take the stored nutrition in the fat to continue functioning. A Riggio Refinement: We could lay-in stores of snacks and beverages; set the alarm for Super Bowl Sunday; take in the game; and, return to our comatose state right after the game, for the remainder of the winter season.

Alas, intelligence, the one trait unique to the self-perceived supreme creature, will never allow a long winter's nap. Our ability to reason, not to mention the use of opposable thumbs, allows us access to the technology required to keep ourselves warm and active, indoors and out, 24 hours a day, seven days a week, during even the most severe South Dakota winter.

Once Homo Erectus discovered fire as a convenient generator of heat, any hope of acquiring the ability to hibernate faded like a fire's last few glowing embers. I believe that moment in history when fire was discovered marked the end of Mother Nature's willingness to comfort our species during winter, as she does the hibernating bear, and the beginning of our awkward attempts to subdue the elements.

Our intelligence elevated the warming process in our homes. From open-face caves, came the need to enclose and insulate. Mud and straw worked well. A fireplace built of clay and stone worked even better. Through the ages, the refinements became increasingly less radical. We still burn things within enclosed spaces, using not only wood, straw, and scatological materials, but natural gas, petrochemicals, and even nuclear reactions. We have the technology to distribute effectively consciousness-sustaining heat to every nook and cranny of our homes. We can also plug in the electric blanket.

Hibernation is no longer an alternative for us, if ever. Anyway, most of us have car payments, rent or mortgage obligations, and so on, due monthly, come summer and winter. Most of us have jobs that will not give us three months off to take a nap. It's just not possible to close our windows and doors for the winter and do nothing but crafts and can fruits and vegetables for the next state fair.

We have a life. Not only a summer life but a winter life, as well. Think of all the great downhill and cross-country skiing, snowboarding and snowmobiling we would miss if hibernation were an option. The bear has no clue what it's missing.

Why Salt and No Pepper?

There is nothing better to warm winter spirits in the Riggio household than a large pot of chili simmering on the stove. Without any scientific basis, it seems that snow falling from the gray skies accumulating on the snow already on the ground accentuates the warm, delicious smells wafting from the kitchen. The aromatic repast emanates from a combination of meat, tomatoes, and beans, slowly cooked, seasoned with chili powder, onion, thyme, red pepper, cumin, salt, black pepper, parsley, a couple of bay leaves, a tablespoon of Jim Dowding's special seasoning and, of course, a touch of sugar. As my dad would always say, "if it is good enough to eat, it is good enough for a little sugar."

Chili warms not only the body, but the soul, and house, as well. So, in my own way, I can't help wondering: if it takes all those ingredients to warm a household, why only salt to warm our streets? Has anyone ever tried cumin or chili powder? Why not sugar or some combination of all the above? Has anyone other than me ever pondered this culinary conundrum, as it applies to our roadways? Alas, probably not.

Trying other compounds for snow removal is not totally deceptive rhetoric. Over the years, road maintenance crews tried many other substances seeking an effective and inexpensive way to clear the streets and highways of ice and snow. Nothing worked as well as good old salt. And, yes, the same salt we shake over french fries and peanuts, can do double duty in removing snow and ice from our roads. The size of the particles best distinguishes road salt from table salt. You can prove it to yourself by grabbing the saltshaker from the table and sprinkling a little on the snow covering the windowsill. It takes only a few minutes for the snow to begin melting.

The unique molecular composition of salt lowers the melting temperature of ice to below 32 degrees Fahrenheit. Lowering the melting temperature causes the snow or ice found on the street and your driveway to return to the liquid state at temperatures lower than the usual freezing point.

Snowplow (Google Images)

How much lower? That exact number depends on several conditions, like traffic, humidity and a few other localized factors. For instance, the higher the volume of traffic, the longer the pavement will remain wet. And, the effectiveness of salt increases when spread on streets just before a significant amount of freezing rain or snow lands on the streets. Proper timing prevents a bond from forming between the ice or snow and the roadway surface. By avoiding that bond from occurring, the plows can more easily scrape away the slippery stuff down to the pavement.

In years past, the Rapid City Street Department uses typically between 10,000 to 16,000 tons of salt per year to keep 850 miles of the city thoroughfares navigable through harsh South Dakota winters. It takes a crew of nearly 20 drivers operating snowplows and/or sanders, around the clock, and often over holidays, if need be, to plow the streets and spread the salt so we can drag ourselves out of bed and into our cars for a skid-free morning ride to work or school.

I, for one, do not take their hard work for granted and truly appreciate their labor. The next time you pass one of these sanders, stay clear, and give the operator a wave and a smile. Probably when you are home during the next major snow event, comfortably enjoying your second helping of chili, I believe the snowplow driver would love to be home, too, instead of seasoning the winter snow with salt.

Winter's Unusual Partially the Usual

Long-range forecasts have improved notably over the years. Some years back, for example, the top 10 climate and weather experts of the Commerce Department's National Oceanic and Atmospheric Administration (NOAA) announced the recent string of record warm winters before the 2000-2001 winter may be over as "normal" winter weather returns. These long-range forecast gurus expected average precipitation and temperatures for the 2000-2001 winter season across the United States.

While dealing with weather as long as I have, I've learned that the abnormal becomes a substantial part of the normal. Winter extremes make up a typical winter, and rarely will a purely normal winter occur without a few days of the abnormal sprinkled in for good measure. So, if we foresaw the 2000-2001 winter season as typical, not only must we appreciate normal, but what about those few days of abnormal weather events used in calculating the normal?

First, "normal," as it pertains to rain/snow, temperature, or wind, conveys the average over 30 consecutive years. Revisions to that value happen at the end of every decade by dropping the first decade of data from the equation and adding the most recent decade of data to the equation, thus keeping the 30-year average intact. Even the extreme or abnormal weather conditions that occur during that most recent decade will be a part of that decade's normal.

A typical winter for Rapid City should have daytime temperatures in the '30s from December through February and nighttime temperatures dropping down into the lower teens. The earliest seasonal subzero temperature occurred on October 27, 1925, when the temperature dropped to 10 degrees Fahrenheit below zero. On the flip side of the coin, the latest seasonal subzero temperature happened on April 3, 1936, with a recording of two below zero. Both events considered outliers and therefore abnormal, were most certainly included in the average formulating the normal temperature.

Normal annual snowfall for Rapid City is about 49.3 inches. Just over six inches during November, just over five inches December through January. Slightly more than six inches during February, nearly 12 inches during March and nine inches during April. Only two inches in March with a trace amount in September. October would produce over two inches of snow. Expect higher amounts of snow in the Northern Black Hills. Lead, for example, can expect 197.5 inches of snow during the winter season. The Southern Hill markedly less. Custer normally receives 55.9 inches of snow. The highest snowfall amounts, on average, happen in March.

Average Annual Snowfall (1981 - 2010)
(National Climatic Data Center)

East River Communities	Annual Snowfall (Inches)	West River Communities	Annual Snowfall (Inches)
Aberdeen	38.4	Custer	55.9
Brookings	31.9	Deadwood	101.2
Huron	43.9	Hot Springs	27.5
Sioux Fall	44.5	Lead	197.5
Vermillion	32.9	Mt Rushmore	52.4
Watertown	35.9	Interior	28.0
Yankton	34.6	Rapid City	49.3
Pierre	31.2	Wind Cave	41.7
Mobridge	31.1	Lemmon	40.6

Extreme winter events occur regularly in and around the Black Hills. One of the most significant monthly snowfall events in Rapid City occurred during April 2013, when 39.5 inches covered the ground. Now keep in mind the average April snowfall is 8.8 inches. One of the most intense snowfall seasons was the 1985-1986 season when 81 inches of the white stuff fell from the skies over Rapid City. Winter storm Atlas dumped more than 30 inches of snow on the Rapid City streets within three days in October 2013.

February 2019 turned out to be one of the coldest and snowiest on record for Rapid City. The airport recorded the lowest average February temperature since 1942. The coldest February occurred in 1936, with a relentless average temperature of 1.4 degrees Fahrenheit. The Rapid City airport logged 16.8 inches of snow ranking the 2019 February the fourth snowiest on record. The normal February snowfall amount for the airport is 6.4 inches.

The least snowfall amount during any one season in Rapid City occurred back during 1940-1941 when a paltry 10 inches fell. You can bet that because these extreme outliers of snowfall occurred and their measurements recorded, those events, both above and below average, became part of the 30-year standard snowfall amounts.

So, there you have it, extreme conditions can occur at any time, and when they do occur, they become a part of the new normal, but they do not represent the normal. When the National Weather Service issues the latest 30-year Rapid City's normal winter temperatures and snowfall in 2020, those average values will include a few nonconforming numbers.

Finally, one or even a few colder than customary seasons or even years does not imply that global warming does not exist. Like establishing those normal winter weather conditions, long-term temperature trends also contain both hot and cold outliers, but the outliers do not represent the temperature trend.

Other Grass Always Greener-Or Whiter

Unlike the fog in Carl Sandburg's poem that arrived on little cat feet, the first winter storm of 2000-2001, like so many other winter storms here in the Black Hills, arrived on not so quiet cat feet. With that one memorable winter storm, came wind gusts over 70 miles per hour, snowdrifts up to one's waist, visibility to less than one-quarter mile due to drifting and blowing snow, and heavy snow across the West River.

As I see it, that particular winter storm brought with it something not so apparent as the in-your-face subzero wind chill. It rekindled the "greener grass" mindset to the folks living in the Black Hills region. With the days leading up to that first winter storm recording relative mild temperatures in the low 60's, the sudden coming of winter felt especially harsh.

It's possible that up until that one storm swept across the Upper Plains, you may have heard a co-worker, family member or neighbor grumble about the lack of winter, better known as the "grass is greener on the other side of the fence." However, with that storm's abrupt arrival, winter returned to West River with much gusto, and many of the "greener grass" folks did an about-face and mourned winter's arrival.

As I see it, our thoughts lament the wicked wintry weather while it is ongoing all around us. But should we really wish for a return to those mild and snowless winter days? After all a winter season with no winter storms means no snowmobile rides, sledding, skiing, snowshoeing, or flowing creeks the following spring?

Now think back when winter came in with its hurricane-force winds whipping snow and cold across West River. I believe it's safe to say that we bemoaned winter when blizzard conditions with below zero wind chills howled outside our door. Just maybe sunny skies were racing through your mind when the snowmelt from your SUV left a wall-to-wall puddle in the garage. Or perhaps you imaginatively recalled those mild, sunny and calm afternoons sitting on the deck, while shoveling the knee-deep snow

Blizzard conditions, Rapid City, SD (Rapid City Journal)

off your driveway. Or possibly the sight of a clear black starry-night sky wandered briefly through your head as you fought the blowing snow while slipping between car and restaurant or movie theater?

Yes indeed, green grass can be found just over the fence, but I think it's green on both sides of the fence. During the dogdays of summer, we'll be anxiously anticipating the first sign of winter. After a period of next winter's bone chilling cold, we will be waiting for the trees to bud suggesting the warmth of spring cannot be far away.

Maybe we should not concern ourselves with the next change in the weather. Enjoy today's weather along with whatever it brings our way. Whether it brings an opportunity to walk in the park or an opportunity to read a book in a warm, comfortable chair while the snow drifts high outside, it's all good.

No matter what "perspective" reminds us about seasons past, the late spring and summer's grass is always greener to better play on and whiter in the winter to better play on, as well.

The Temps Might be Low but it's No Ice Age

A few weeks before the start of the 2000-2001 winter season, the temperature at the Rapid City Regional Airport fell below zero for the first time in about 705 days or just shy of two years. Possibly due to this prolonged absence of subzero temperatures, that week's frigid temperatures felt like the onset of the next Northern Plains ice age.

In our Earth's 4.5-billion-year history, the planet had experienced several periods when glaciers covered a significant portion of the planet's surface. At their peak, the thick ice sheets covered up to 30 percent of the Earth's surface, with each ice period lasting tens of millions of years.

The most recent ice age lasted from about 1.6 million years ago to about 10,000 years before present. At least 20 times during that period, the icy expanse would grow then retreat, only to expand again. The most recent ice growth spurt in North America, known as the "Wisconsin glaciation," hit its peak about 20,000 years ago.

The climate temperature during that particular ice age registered as much as 27 degrees Fahrenheit colder than temperatures today. Despite that recent ice age cold snap, the past 10,000 years have been a part of a relatively warm interglacial period. Interestingly, the presence of massive continental ice sheets on Greenland and Antarctica, along with numerous smaller glaciers in mountainous regions throughout the world, indicate that the Earth could still be in the grip of an ice age.

Because no records exist regarding any of the ice ages, the only evidence of their existence comes to us from the land and sea environments. Solid rocks, and not the ice itself, first suggested the presence of the ice ages.

Scientists surmised that the large smooth rocks found scattered across meadows and forests seemed totally out of place when compared with surrounding cliffs. If their presence came not from the nearby cliffs, then maybe something carried the rocks in from an outside source. To some scientists, the biblical story about Noah's flood was the most apparent means of transport for these boulders.

However, two early 19th century scientists concluded that the misplaced boulders found in many of Switzerland's valleys originated high in the Alps and were carried down from the mountains by large advancing glacier fields. In 1837 this ice age theory proposed that an enormous sheet of ice moved down from the North Pole covering not only Eurasia but North America, as well.

They proved their theory by showing that these gigantic sheets of ice changed the shape of the land, decreased sea level heights and even resulted in scattering across the continents many land-locked lakes. They showed valley walls and floors scratched, grooved and polished by slow-moving glaciers. The U.S. Midwest originated from a slow-moving sheet of ice, possibly more than a mile thick. Today's Great Salt Lake happens to be the aftermath of a retreating glacier.

What triggered the ice growth and, more importantly, what caused the ice to retreat? Once sheets of ice begin to grow and cover more and more land, the expanding ice sheets reflect more and more solar energy into space, prompting the climate to cool further. A colder climate means further growth of the ice sheets, which in turn cools the environment even more.

Several theories of what might have triggered climate changes come to mind. One speculation concludes that volcanic eruptions filled the sky with enormous amounts of soot and cinders blocking the sun's energy from warming the Earth. The year 2000's Jasper fire gives us only a tiny glimpse of how smoke can effectively reduce sunlight from reaching the Earth.

Another theory addresses the uplifting of continental blocks. Plate movements cause an uplift of towering mountain ranges, which in turn can cause profound changes in the global oceanic and atmospheric circulation patterns. In some respects, the uplift of the Himalayas and the Tibetan Plateau contribute to today's world climate.

Others speculate that a reduction of the Earth's atmospheric heat-retaining gas, carbon dioxide, contributed to the development of the ice ages. Many processes can cause a long-term decrease in the amount of atmospheric carbon dioxide. Chemical reactions caused by interactions among organisms, ocean currents, erosion, and volcanism can reduce carbon dioxide. Still, uncertainty exists regarding the magnitude of past carbon dioxide changes to be large enough to initiate and sustain the ice ages.

Finally, changes in the Earth's orbit can have a profound impact on the Earth's climate. Pivotal orbital parameters known to vary over many years include the eccentricity or stretching of the orbit around the sun and the tilt of the Earth's axis. The elliptical orbit of the Earth elongates from 1 to 5 percent through time. With a highly elliptical orbit, one hemisphere will experience much warmer summers and colder winters. A nearly circular orbit guarantees similar seasonal contrasts in temperatures in both hemispheres. The Earth's orbit goes through one of these stretching cycles approximately every 100,000 years.

The Earth's tilt offers up yet another variation. Today the Earth's axis tilts approximately 23.5 degrees. The tilt wobbles from 21.6 degrees to 24.5 degrees roughly every 41,000 years. Changes in the angle of the Earth's axis can cause substantial changes in the seasonal distribution of radiation at the higher latitudes, lengthening the winter's dark period at the poles.

Together these two Earth-bound variations combined can cause radiation changes of up to 15 percent at the high latitudes. Changes of this magnitude of sunlight reaching the Earth's surface can significantly influence the development, growth, and melting of ice sheets.

The ice age may not be a product of any one of these theories, but possibly some combination of each with an unknown thrown in for good measure. Before the human-induced global warming of the last two centuries, evidence suggests a naturally cooling worldwide climate for the past several thousand years.

How Cold was the Ninth Coldest Winter?

The Rapid City National Weather Service, whose job it is to track daily temperatures, reported, a few years back, that the 2000-2001 winter season was the ninth coldest on record in Rapid City. Considering that their weather records go back well over a hundred years, ending up ninth should be noteworthy.

The long-range forecast predicted a cooler 2000-2001 winter. Sure enough, daily average temperatures during the 2000-2001 winter slipped dramatically down from the previous winter's mild levels and fell well past the average range and into the top ten of below-normal winters.

The simple average between a day's high and low temperatures assigns that day's daily average temperature. From November through February, the overall 2000-2001 winter season average temperature was 22.7 degrees Fahrenheit. By ranking this seasonal temperature with previous winters' seasonal temperatures, it ended up in ninth place of the coldest winters. In first place, or the coldest average temperature for that same period was 17.5 degrees Fahrenheit during the winter of 1978-1979.

Without the statistical hoopla, can we in good conscience award the 2000-2001 winter the title of ninth coldest winter? Think about all the historical blizzards that blew through the Dakotas. The 1886-1887 blizzard that wiped out the most celebrated herd of range cattle in the world. The school children's blizzard of 1888. The long winter of 1906-1907 when the weather kept bringing Arctic cold and blizzard snow. The bone-chilling temperatures during the winter of 1932-1933. The more recent blizzards of 1949, 1951-1952, 1966, 1971-1972, 2000, 2013, and 2019.

With these past frigid winters as our backdrop, the ninth coldest winter should conjure up memories of people-bearing cocoons of warm clothing slowly moving about the city and automobiles with snow piled high on their roofs barely moving through the not so white canyons of snow.

Nothing of that sort happened during the ninth coldest winter in our history. Of course, we had our cold days that, on occasion, turned into a cold week or a cold two weeks, but a warming trend typically followed those weeks. After all, the high temperature on the last day of the 2000-2001 winter was 60 degrees Fahrenheit.

We have become accustomed to mild winters here in West River thanks to the westerly downslope winds, or chinook winds coming off the Black Hills. At worse, the warming winds made the ninth coldest winter tolerable, especially when compared to other areas of the state and country. Farther east beyond the Missouri River Valley and out of reach of the chinook wind effects, that winter did indeed feel more like the ninth coldest on record.

East River and most of the Midwest experienced the worst of this now long forgotten winter. Our South Dakotan neighbors in East River experienced a hard, cold 2000-2001 winter with record snow. In no small measure to all the snow on the ground east of the Missouri river, temperatures there remained bitterly cold. Huron, SD documented record snowfall. In the heartland, Des Moines, IA recorded more than one hundred days with one inch or more of snow on the ground. Chicago, Il too suffered through a bitter winter.

The winter of 2000-2001 was colder than usual, not only in the Midwest but across the lower 48 states, sliding into the 26th slot for the coldest U.S. winter on record, a distinct contrast to the previous two warmer than average winters in the U. S.

On a global perspective, the Northern Hemisphere winter season temperatures were contrary to the U.S. winter season temperatures. Warmer than normal temperatures were recorded across the northern half of the globe making the 2000-2001 winter the seventh warmest winter period on record across the world.

So how cold is the ninth coldest winter? It does not matter. In truth, a cold day or winter to some may not be the same to others. Cold exists in the viewpoint and senses of the beholder. To me, cold is not so much a function of temperature but based on whether I can golf or go skiing. If I can play golf, it's not a cold day, despite what the daily average daily temperature says. If I can ski more days than not during a winter season, it's also not a terribly cold winter.

With A Cold and Snowy Winter Come the Floods

The 2000-2001 winter etched several South Dakota snowfalls records into the books. Total snowfall exceeded 80 inches in Huron — a new best. The number of consecutive days with an inch or more of snow on the ground tied or exceeded long-time records in many East River communities. In brief, South Dakota — and much of the Upper Plains — got more than just the typical cold and snow that long-ago 2000-2001 winter.

The aftereffect of a hard, cold and snowy winter endured well into spring. The snow persisted a little longer, the ground remained frozen a little longer, and the rivers and streams stayed iced over a little longer. Pushing all these ingredients into spring often equates to damaging floods, particularly in East River.

With record snow on the ground across East River, and increasingly warmer temperatures and numbers of wet storm systems migrating across the state, it stands to reason that some of the state's major river basins would flood. The snowmelt rushes across the hard-frozen ground toward the ice-jammed tributaries and rivers, while the early spring rains add additional water to the landscape that too quickly flows into the James, Vermillion and Big Sioux river basins. Soon, farmlands and river towns such as Watertown, Mitchell, Forestburg, Scotland, and Huron found themselves battling advancing floodwaters.

Huron recorded a James River flood stage at just over 18 feet, about seven feet above bank full. At Watertown, Lake Kampeska stabilized just shy of 4 feet above flood stage. Lake Poinsett reached 18 inches above bank full.

The topography of these East River basins in no small measure keeps them flood prone. The gradual slope of the land throughout much of Eastern South Dakota willingly allows floodwaters, albeit shallow, to spread great distances away from the riverbanks.

Ft. Pierre, SD, during the 1952 flood (Yahoo Images)

Some of the more noteworthy East River floods over the past few decades include the 1952 flood. Then, the rapid melt of an above-normal snow cover in Eastern Montana, North Dakota, and South Dakota caused record flooding along several tributaries to the Missouri River in North and South Dakota, James River and the Big Sioux River. Anchorman Walter Cronkite credited the Weather Bureau's river forecasting service — "a fantastic modern weapon against flood dangers" — with saving the lives of hundreds of people by forecasting this flood ten days before the crest reached Pierre.

Large areas of the Missouri River basin received intermittent heavy rainstorms during May and June 1984. The prolonged wet spring culminated in June with the worst flooding since the disastrous flooding in 1952. The tributary river basins in which flooding or high flows occurred were again the Vermillion River, James River, and the Big Sioux River basins, along with watersheds in Iowa and Nebraska.

During the spring of 1995, unseasonably warm temperatures in February and March resulted in quick snowmelt that saturated the ground. These warm temperatures, along with repeated heavy rains in the Midwest during March, April and May, produced significant flooding on many streams and rivers in the Dakotas, Nebraska, Iowa, Kansas, and Missouri. By mid-May, the James and Big Sioux Rivers, and their tributaries overran their banks and thoughts of a repeat of "the great flood" swept through people's minds. The same flooding scenario repeated itself during the Spring of 2019.

The Missouri Basin River Forecast Center located in Pleasant Hill MO, just southeast of Kansas City, issues river forecasts for the James, Vermillion and Big Sioux river basins. Like any prediction of Mother Nature's whim and fancy, small changes in rainfall amount and location, snowmelt, river ice, topographic changes and the amount of debris floating down the river, impact the accuracy and timing of the river forecast. Consequently, river forecasts, like all weather forecasts, maintain an inherent degree of uncertainty, and therefore, all users of these river forecasts, or any weather forecast, must assume a degree of risk.

A Poet's View of Winter Solstice

When Robert Frost put pen to paper and wrote one of his best-known poems, "Stopping by Woods on a Snowy Evening," he described the first day of the winter season precisely. In the second stanza, he writes, "My little horse must think it queer/To stop without a farmhouse near/Between the woods and frozen lake/The darkest evening of the year." Undoubtedly, his ride into the woods to enjoy the solitude and beauty of the nighttime woodland took place during the winter solstice – the darkest evening of the year.

A long winter's night (Dan Lutz)

I believe his stanza "The darkest evening of the year" refers to the maximum number of hours of darkness across the Northern Hemisphere, which occur on the 24-hour period of the winter solstice. On this day, the sun travels the year's longest path through Earth's atmosphere to reach Frost's hoof-marked path severely lessening its warming energy, keeping the lake near the woods froze solid.

Ironically, Earth finds itself located nearer to the sun during the winter solstice than during the summer solstice – by about 3 million miles. Regardless of Earth's proximity to the sun, the 23.5-degree tilt of Earth causes the cold of the winter and the warmth of the summer. You see, the angle determines the number of hours and minutes of the warming sunlight and chilling darkness.

On the day of the winter solstice, the sun will find itself at its farthest point from the equator directly over the imaginary line circling the Southern Hemisphere known as the Tropic of Capricorn. The noontime elevation of the sun will appear to remain the same for a few days before and after the solstice, hence the meaning of the Latin term "solstice" – "sol" meaning "sun" and "stitium" meaning "a stoppage."

Many ancient pagan civilizations feared that the failing life-giving light from the sun might eventually drop below the horizon and never return. These civilizations believed that only through human watchfulness and celebration would the sun pivot away from its failing state in the heavens toward a rebirth of its life-giving presence. These celebrations conveyed love to family and friends, feasting together, the giving of gifts and the lighting of trees and the yule log to keep the growing darkness at bay.

It's supposed that the Christmas celebration has close ties to the winter solstice celebrations. December 25 was the date of the winter solstice in the Roman calendar about 46 B.C. Today, because the Roman calendar did not match precisely the length of the solar year, the winter solstice occurs on or about December 21.

Christians believed that any celebration of the sun, in fact, celebrated Christ, who conquered darkness. About 273 A.D. – a few decades after Emperor Constantine's conversion to Christianity – the celebration of the birth of Christ shifted from Three King's Day on January 6 to the ancient Roman-calendar winter solstices on December 25.

After the winter solstice, the days begin to grow longer, permitting more work to get done before dark. I believe Frost had this puritan work ethic in mind when he wrote in the final stanza of his poem, "The woods are lovely, dark and deep/But I have promises to keep/And miles to go before I sleep/And miles to go before I sleep."

Dan Lutz

The weather has often played a leading role in the history of this country, whether in the mind of my grandfather or in the spirit of Gen. Dwight Eisenhower on the eve of D-Day.

🍃 PAST WEATHER EVENTS

My first awareness of weather's impact on history came in the basement of my childhood home in Chicago, IL. There, my grandfather and I often discussed a host of subjects. Many of those topics, such as his early days in the Army, always included his remembrance of weather, which stood fast in my mind over the years.

My grandfather, like many older men of the day, could make his point using very few words. It seemed his economy of speech — slowly delivered, well-chosen words — left vivid images in my mind of places he'd seen outside those basement walls where he found himself on many a cold winter day.

He told me about his time in the military when he trained at Fort Meade, SD. He recalled a view of a "big butte" visible through the window of the army barracks. I now know that he described Bear Butte. Back then, the Cavalry bivouacked closer to Bear Butte than the current location of Fort Meade Veterans Administrative Hospital. The Centennial Trail traverses the old Cavalry grounds near the base of Bear Butte, near the site where my grandfather camped.

The weather has often played a leading role in the history of this country, whether in the mind of my grandfather or in the spirit of Gen. Dwight Eisenhower on the eve of D-Day. My grandfather described magnificent rainstorms that sprang up suddenly over the grass prairies, and how quickly they moved across the campgrounds. He knew there was nothing anyone could do about the storm and the mud to follow, except to shout back defiantly into the winds: "Let it rain. We were here first!"

Little did my grandfather know that the Army did try to control the weather.

War and Rainmaking

Around 1871, military historian Edward Powers, an expert on Napoleon's battles and observer of the Civil War, contended that heavy artillery bombardment caused increased rainfall. Powers came up with his hypothesis following many long talks with Civil War veterans. Powers took careful notes of the stories the veterans recalled about how, inevitably after intense cannonading, a subsequent rain would end the battle for the day.

With the publication by Powers of the book, "War and Weather, or Artificial Production of Rain," he quickly became the leading proponent of a novel rain-enhancement theory. The theory stated that the continuous cannon fire created concussion waves. The waves somehow gathered and combined many small clouds from throughout the wide battlefield area and brought them together to form more massive clouds over a smaller area. Eventually, these more massive, stronger clouds produced rain.

Powers lobbied for funds to test his hypothesis, and after years of Powers' persistence, Congress appropriated $10,000 for this far-reaching experiment. That, plus another $2,000 from the Department of Agriculture, allowed Gen. R.G. Dyenforth of the U.S. Army, to carry-out rainmaking experiments based on Powers' ideas.

West Texas rainmakers (Google Images)

Gen. Dyenforth assembled a scientific party of 13, including a meteorologist, a balloonist, an electrician, a statistician, a chemist, and Professor Powers himself. The team conducted its rainmaking experiments for two years near Midland, TX. The tests involved sending gunpowder aboard gas-filled balloons, and detonating explosions aloft, thus simulating the rainmaking canon fire.

The experiments ended inconclusively, causing the research team and the theory to endure considerable public criticism and ridicule, especially after an unflattering article published in "Scientific American." After the article was published, Gen. Dyenforth was unfortunately labeled Gen. "Dry Hence Forth."

A Dynamite Rainmaking Theory

Attempts to make it rain did not end at the turn of the 20th century, even after inconclusive experiments by Army Gen. R.G. Dyenforth. Gen. Dyenforth attempted to prove or disprove the Powers' theory that loud explosions in the air would gather clouds and make them rain. The quest to turn parched deserts into productive cropland buoyed the concussion theory for many years. The concept particularly appealed to one of America's most renowned visionaries, Charles William Post.

C. W. Post, born in 1854 in Springfield, Il, grew up amid an agricultural background. He used his inventiveness, industriousness, and talents as an entrepreneur to make a vast fortune. During the last two decades of the 18th century alone, Post received patents that improved seed planters, plows, cultivators and hay stackers. He also patented cooking utensils and even suspenders. Thanks to Mr. Post, bicycles today have two of the same-sized wheels.

Eventually, the pace of success and the burdens imposed by looking after his fortune took their toll. Post had a severe breakdown and physical collapse. His doctor advised a move to the warmer and drier climate of West Texas. For a while, C.W. Post enjoyed the pure pleasure of roaming the land as a cowboy. It was not long before his penchant for entrepreneurship returned to launch what became a thriving cattle business.

One day, while resting under what shade a mesquite tree afforded, gazing out from his vantage point high on the West Texas Caprock, Post envisioned a beautiful city situated in an oasis amid the desert expanse before him. Unfortunately, Post's health failed him again, and he boarded a train for a famed sanitarium located in Battle Creek, MI. Gradually, he regained his health with a regimen of diet, mental science, calisthenics, and water therapy.

Rejuvenated and impressed with the changes brought on by his newfound healthy eating habits, Post remembered a mixture of chicory, roasted wheat, and other grains often served on the Texas Plains as an alternative to coffee. He began trying to develop a substitute coffee that had the same taste but without the caffeine.

A year of relentless work and experimentation yielded not the coffee substitute, but Postum cereal. Later, Post developed Post Toasties and Grape Nuts. As the cereal company blossomed into one of America's largest industrial corporations, Post again suffered the debilitating by-product of his success. Once more, he was in pain and utterly exhausted from stress and workload. So, Post returned to West Texas for rest, relaxation and to tinker with his dream of an oasis in the desert. If he can make it rain in the West Texas desert, he thought, he can grow wheat and oats as far as the eye can see for his ever-expanding cereal business.

Post's experimentations already had made him a wealthy and famous man, but nothing he did previously came close to matching the scope of his next project. Based on Powers' rainmaking theory, Post's first operational experiment to make rain began in 1910. Workers attached dynamite to huge kites, flew the kites toward clouds, and ignited the explosives. It took only one such test to prove that the kite detonation method was too dangerous to pursue.

So, later, Post located 14-pound packs of dynamite every 50 feet for a quarter of a mile along the ground, and at the appropriate set of weather conditions, he set off the blasts in 10-minute intervals. During one such battle, as Post referred to them, he set off 3,000 pounds of dynamite. Rain fell almost immediately.

Post exploded more than 24,000 pounds of dynamite in 1912 alone but saw only a slight rainfall in the Texas communities of Crosbyton, Slayton and Post City. The experiments continued into 1913 but soon ended when the rain became plentiful naturally.

Operational rain-enhancement commercial projects and experiments continued in West Texas, but the land remains a desert. Still, we must allow forward thinkers like C.W. Post to go beyond the obvious in search of their elusive dreams. If we do, Post's vision gleaned from underneath that mesquite tree, of turning the scorched desert into fertile cropland, may yet become a reality.

Meteorologists Make Rainmaking Discovery

Long after attempts by military historian Edward Powers and cereal maker C. W. Post to make rain, a meteorologist made a fundamental discovery. While working at the General Electric Laboratories in upstate New York, meteorologist Vincent Schaefer discovered that man might be able to change the dynamics of a cloud to bring about rain. Working with a modified freezer, Schaefer created an artificial cloud, like the cold mist that rushes out of your household freezer when you open the door looking for ice cream on a hot day.

Schaefer next placed a chunk of dry ice into the freezer to keep the artificial cloud well below freezing temperatures. Immediately, the cloud's supercooled water droplets (pure water may exist in the liquid form at temperatures as low as minus-40 degrees Fahrenheit) transformed into ice crystals, releasing the latent heat of fusion.

Vincent Schaefer with the artificial cloud chamber (Google Images)

The observation made Schaefer realize how the dynamic characteristics of a cloud could change in a way that would benefit humankind. You see, cirrus clouds found in the upper reaches of the atmosphere are made up of ice crystals. The same holds with most rain-producing clouds, whether the big, black, lightning-producing cumulonimbus clouds of summer or the flat, gray nimbostratus clouds of winter. As they grow into the much colder upper atmosphere, billions upon billions of these small water drops, that make up the cloud's beginnings, become supercooled, i.e., remaining liquid at well below freezing temperatures. By transforming the supercooled water into ice crystals — snowflakes — they can quickly grow to larger sizes at the expense of the remaining supercooled water from surrounding small droplets. Reason: the saturation water vapor pressure is higher than the saturation ice vapor pressure. This difference in vapor pressure allows the water vapor to travel from high vapor pressure (water) to low vapor pressure (ice).

The ice crystals typically grow large enough to fall and melt in the warmer temperatures in the lower atmosphere, becoming rain. This is precisely how nature forms most of the rain that falls anywhere on the globe. Also, the latent heat released from this water-to-ice metamorphosis adds to the buoyancy of the cloud, allowing it to grow dynamically, forming a more efficient rainstorm.

Schaefer imagined "seeding" clouds which contained super-cooled water droplets with dry ice to grow the ice crystals large enough to fall out as rain. His excitement led to experiments high over the plains of New York in a specially equipped airplane, dropping dry ice pellets into the expansive layer of a supercooled stratus cloud. The research contained a touch of whimsy: the plane delivered the ice crystals in a pattern that spelled out the initials, G. E.

In 1947, the General Electric researchers hypothesized that, if they could change the characteristics of clouds, they could alter the characteristics of hurricanes, whose muscle comes from clouds. The idea would be to modify a hurricane's intensity by inducing an early release of some of the tremendous energy stored in the maturing storm.

On October 13, 1947, a hurricane appeared 400 miles off the Florida coast, heading out to sea on a northeasterly track. Two aircraft flew over the storm and dropped dry ice into as many clouds as their fuel, and dry ice supplies would allow. The researchers noted but did not document photographically, that the structure of the storm changed. Unfortunately, a few hours later, the storm reversed course and headed west toward land.

The hurricane and all its accompanying fury made landfall on the Georgia-South Carolina beaches. After that, the suggestion of human intervention into the ways of nature opened all manner of political and legal issues that severely dampened further hurricane or rain modification experiments.

Behind the D-Day Weather Forecast

If you have one timepiece, you will always know the time. If you have two watches, you will never know the time. The same philosophy holds for weather forecasts. With one weather forecaster, you know the forecast. Having more than one forecaster, you never know the forecast. Now, try to imagine preparing one of the most critical and famous weather forecasts ever in the history of this country, with a multitude of forecasters.

A weather forecast for the largest fleet ever assembled in its time, to transport 176,000 troops on more than 5,000 ships and landing craft, with air support of over 8,500 aircraft. Should you use one forecaster or the combined knowledge of many forecasters?

This was the plight of Gen. Dwight David Eisenhower, the supreme commander of the projected Allied invasion of Northern France. Gen. Eisenhower fully understood that successful execution of the initial goals of Operation Overlord depended extensively on a detailed and accurate D-Day weather forecast. He not only wanted to comprehend the meteorology behind the D-Day forecast entirely but, more importantly, he also wanted to understand the integrity of the forecasters preparing the weather prediction.

"I know little about the science of meteorology and certainly nothing about what reliance can be put on it for forecasting in England, but, as the day will soon come when a weather forecast may be a critical factor in an important decision which I shall have to make, I want the first-hand experience not just of the forecasts. I want to know my meteorological advisers and what they can do. I want to know when and how far I can trust them," he said.

The supreme commander counseled with two meteorologists: British group Capt. James Martin Stagg from the British Meteorological Office, who held the senior post; and U.S. Col. Don Yates, who held the deputy post. Incredibly, these two gentlemen mediated professional opinions from, not one, nor two, but three weather centers, each staffed by numerous meteorologists, to arrive at the one historical weather forecast.

The three centers were established at different places to prevent a single German attack from wiping out all the D-Day meteorologists. At least that was the official explanation. Unofficially, decentralizing the three centers into independent forecast centers had more to do with the fact that if the D-Day forecast were inaccurate, no one national group — British or American — would shoulder the blame.

One center was the U.S. Strategic Air Force central (code name Widewing), located within walking distance of Gen. Eisenhower's headquarters. Col. Yates commanded Widewing. Two Americans, Ben Holzman and the outspoken Irving P. Krick, supervised four small, long-range forecast units. Each of these small groups would use different methods to produce five- to seven-day forecasts.

One method would find a historical weather pattern that reasonably well-matched the current weather pattern. Then the long-range forecast was prepared based on past events, following the historically matched pattern. The other groups would use methods adopted by Eastern Airlines statistical techniques, and even methods developed by the Germans.

A second center included the Royal Air Force's (RAF) civilian operated weather group located in the London suburb of Dunstable. Heading this team were the dean of the D-Day forecasters and distinguished World War I RAF pilot and weather observer, Mr. C. K. M. Douglas, and the Norwegian synoptic meteorologist Sverre Petterssen.

Douglas used his uncanny ability to recall unique weather conditions from his past to predict future weather changes. He often expressed little faith in the "modern" methods asserted by the American foreigners operating out of Widewing.

The third group included the forecasters from the Royal Navy. They were located two floors underground in the Admiralty in downtown London. This group was made up of mostly British engineers, mathematicians, and naval meteorologists.

Difficulties and intellectual arguments exposed their ugly selves on many different levels. For one, trying to get one agreed-upon forecast from three self-governed weather centers were near impossible. Secondly, the experience, personality, and disposition of the principals managing each of the three weather centers ranged the entire psychological and professional spectrum. Disagreements were the rule, not the exception.

It would appear to the casual observer that the only correct prediction from this gamut of meteorologists charged with preparing the D-Day weather forecast would be chaos. Thank goodness it wasn't.

Weather requirements for bombers, gliders, paratroopers, landing craft, naval forces and ground forces exist at opposite ends of the weather spectrum. Bomber pilots need clear skies to spot their targets better, while paratroopers covet cloudy skies to protect them from enemy weapons. Troop amphibious landing craft want onshore winds to reduce wave action on the beach, while the larger naval ships prefer offshore winds with the resultant small waves.

Also, the D-Day invasion of Normandy, France had to occur during low tide to uncover as many water obstacles as possible. Further, the landing force needed three subsequent days of good weather to reinforce and re-supply the ground troops. This one forecast was monumental in scope and impact.

This awesome responsibility rested square on the shoulders of Group Captain and British meteorologist James Martin Stagg, his deputy, Col. Yates, and the three weather centers under their command.

After much discussion among the commanders of the different invasion forces, Gen. Eisenhower established just what the weather conditions must be to satisfy the needs of all the attacking armed forces. The minimum D-Day weather conditions must include: a full moon within one day before four days after the invasion; near calm seas; visibility 3 miles or greater; cloud cover less than three tenths below 8,000 feet, with cloud bases above 3,000 feet. Historical weather records concluded that only a slim probability, at best, existed of ever seeing these desired weather conditions on the beaches of Northern France during June, maybe one day out of 13.

Starting on April 17, 1944, meteorologist Stagg would brief Gen. Eisenhower and the air, ground and naval commanders-in-chief and their staff every Monday on the week's weather. For the weekly briefings, Thursday was always D-Day. The following Monday, Stagg critiqued the previous week's forecast. Ike also instructed Stagg to inform him immediately "whenever you see a good spell that would be suitable for Overlord coming in the next month or so, give us as much notice as possible."

On May 8, 1944, Eisenhower set June 5th as D-Day. So now Stagg had the one day to target his forecast. As May's more stable weather began its transition into June's more unstable weather, the large-scale weather patterns began to break down. On June 1 a polar front stretched from the Aleutians across the U.S. and the Atlantic to Europe, with a series of fast eastward-moving wave-like storm systems sliding all along the front. It appeared that the key ingredient to a successful June forecast rested on predicting the strength and movement of the Azores high-pressure system over the central Atlantic.

The ebb and flow of this high-pressure system would dictate the direction of movement of those eastward-moving storm systems relative to the coast of Northern France.

On June 2, the forecast center at Widewing (Krick and Holzman) believed that by June 5 the Azores high would direct the storms away from Northern France, therefore making the weather suitable for the invasion. Petterssen and Douglas at Dunstable, on the other hand, believed just the opposite: the weather conditions would be unsuitable. The Admiralty forecast favored the Widewing outlook. Tempers flared among the forecast centers, with Stagg in the eye of the storm.

With less than an hour before Stagg had to brief Ike on an "agreed to" forecast that would cover the next five days, he decided to go against Widewing, his deputy Yates, and the Admiralty. Stagg briefed the commanders that the outlook for June 5 was not good.

The next morning, Saturday, June 3, the dichotomy of forecasts persisted. The three weather centers did not budge from their previous day's forecast; however, that evening's analyses caused both the Widewing and Admiralty forecast groups to acquiesce their heretofore upbeat outlook partially. They ever so reluctantly joined in Dunstable's more pessimistic June 5 forecast.

Stagg finally had an "agreed to" forecast. With current outside conditions clear and calm, he again briefed the commanders-in-chief that the passage of a powerful cold front across the channel would bring gale-force winds by late Sunday, June 4, and early Monday, June 5, along with nearly overcast clouds, low ceilings, and stormy conditions.

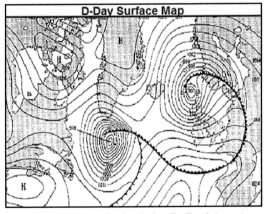

The weather map of the day before D-Day's invasion, June 5, 1944 (Google Images)

Based solely on the forecast, and over the objections of British commander Bernard Montgomery, Ike postponed D-Day 24 hours. Ike reasoned that the weather prediction would prevent Allied air superiority, the most crucial element to a successful invasion, given the slim margin of Allied ground superiority.

A reversal of fate would fall upon Stagg. Weather analyses completed the following day now showed changes in the weather patterns that would push the cold front through the channel faster than expected, leaving a brief period of improved weather in its wake. With all three forecast centers in agreement, and as the winds increased in strength under overcast skies heavy with rain, Ike heard a much-improved forecast that Sunday evening.

The forecast called for 50 percent or less cloud cover with bases at 2,000 feet to 3,000 feet as well as decreasing winds from Monday afternoon until at least dawn on Tuesday, June 6, or possibly longer. In short, the forecast called for less than ideal conditions for a June 6 assault, but tolerable. After consulting with his staff and commanders, Ike said, "I don't like it but, there it is… I don't see how we can do anything else. I say we go." D-Day was set to go for early June 6.

It is hard to imagine the enormity of the D-Day operation, let alone the importance of that day in the history of the world. The pressure to deliver an accurate forecast was extreme. Fortunately, or unfortunately, the weather for D-Day was fundamentally as forecast: marginally operational. More than 176,000 troops on 5,000 ships and landing craft, with 8,722 aircraft overhead, were set in motion on that day — come hell or high water. Sadly, the invading troops encountered both hell and high water off the coast of Northern France.

The first phase opened just after midnight with an airdrop of men and equipment over Caen, France. Dense clouds prevented many of the paratroopers from finding their drop areas, and 60 percent of the gear was lost. Operation Overlord began somewhat ominously due to poor weather.

The winds over the channel produced blinding spray and cold rain that lashed the decks of the landing craft, causing moderately choppy seas with waves of 5 feet to 6 feet at mid-channel, and 3 feet to 4 feet in the embarking area about 10 miles offshore.

Black Hills resident Wayne Brewster, who landed on Omaha Beach a few minutes after H-hour (the hour of the attack, 6:30 AM), recalls those terrible sea conditions at the embarking area. "Myself and Corp. Stacovich were delegated to be the first to go down the landing net to our LCVP. We were combat engineers and heavily loaded. Besides my field pack, my load [included] 25 pounds [of] TNT. I was one of the smallest guys, so my load was light. We were ordered on first because of the heavy seas — still huge waves because of the earlier storm — to help the other men off the landing net to avoid injuries, if possible. If someone tried to get off the net as the boat (LCVP) was dropping, serious injury [could happen] and if the boat (LCVP) was coming up, the same could happen. It goes without saying that we all were seasick."

Eisenhower reported later, "The high seas added enormously to our difficulties getting ashore. Landing crafts were hurled onto the beaches by the waves, and many of the smaller ones were swamped before they could touch down. Numbers of troops were swept off their feet while wading through the breakers and were drowned, and those who reached the dry land arrived near exhaustion." The much less-than-ideal weather contributed mostly to the emotional make-up of now exhausted, seasick and tense soldiers who hit the beaches of Northern France, all the while under heavy enemy fire.

The breaks in the clouds were spaced far enough to allow naval bombardment but proved to be more difficult for air operations. The cloud deck split into two layers, with a ceiling of 10,000 feet. At H-hour, the beachhead had variable cloudiness. At Omaha Beach, the heavy bombers could only attack enemy fortifications blindly through overcast clouds. Some degree of aerial success occurred at Utah Beach where cloud breaks were more prevalent. Generally, the most significant contribution made by Allied bombers was mostly psychological — depressing the German defenders, while offering greater confidence to the Allied ground troops.

Despite the adverse weather conditions, Sverre Petterssen stated that the D-Day forecast was, "Meteorology's finest hour." British group captain James Martin Stagg, the senior forecaster, reflected that consulting with three independent weather forecast offices to arrive at the one forecast was the right choice. U.S. officer Col. Don Yates, who held the deputy post to Stagg, disagreed.

In summary, the Allied forecasters saw the slight break in the weather coming, albeit slight, that allowed the D-Day invasion. The German meteorologist did not. Consequently, the German defenders were caught somewhat by surprise, thinking the forecast weather conditions would delay an attack. In January 1961, while riding with John F. Kennedy to his inauguration, Ike told JFK that the "Allies probably prevailed because of superior meteorologists. German meteorologists, however, were not inferior, the data to them was."

Dust Bowl Drought: Nation's Greatest Weather Disaster

One of, if not the most devastating weather disaster in American history, was the drought during the decade-long Dust Bowl era of the 1930s. Although the Southern Plains felt the drought's full vengeance, the windblown dust and agricultural decline were no strangers to the Northern Plains.

Excerpts from diaries tell the South Dakota story. June Brindel wrote, "There were few trees around the South Dakota farm, and the wind keeps blowing. Dust blew all the time. I remember the wild pheasants got so hungry that they came right up to the house looking for food. My brother would come and grab them, and we'd have pheasant for dinner. It was survival of the fittest."

Dust Bowl conditions in Dallas, SD (Google Images)

She also recalled seeing a dark cloud in the west, thinking rain was on the way. Unfortunately, the cloud was not a rain cloud, but a horde of grasshoppers — millions and millions of voracious grasshoppers looking for anything green to devour, including the paint off buildings.

It is interesting to note that the disastrous impact of the Dust Bowl drought began decades before the last soaking rain fell from the heavens onto the Northern Plains. The Dust Bowl started with the Homestead Act of 1862, offering free land to those willing to farm it across the fertile green Midwestern and Northern Prairies. Almost overnight, millions of acres of grassland were plowed under to make room for wheat. The settlers soon discovered, however, that while these vast grasslands were productive in wet years, they also were subject to severe drought.

Occasionally, excessive dry periods caused the crop to fail, forcing farmers to vacate their homestead and move on to greener areas. The abandoned prairie never had a chance to recover its natural vegetation fully because the plows returned with the rains on the Midwest and Northern Plains. Around 1920, tractors replaced horses, allowing farmers to plow under even higher amounts of grassland. More and more of the once lush green prairie was laid bare, anxiously awaiting the next rain.

During the late '20s and very early '30s, life on the farm was excellent, with rainfall amounts about 5 inches above normal. Not such a good experience on Wall Street, however. The bottom dropped out of the stock market, and the Great Depression spread into the Midwest and across the country. The price of cereal plummeted, forcing the Midwest and Northern farmers to cultivate even more prairie land with the hope of growing more grain to make ends meet. By 1933, the Midwest and Northern farmer literally bet their sizeable farms on the next soaking rain — rain that would never come until nine dusty, dirty years later.

In 1933, the dice tossed for the farmer's weather came up snake eyes. The typical active rain pattern changed into an unmoving pattern void of moisture. Warm and moist winds from the Gulf of Mexico were diverted to other areas of the country and away from the Plains states. Year after year, the moisture-bearing westerly winds blowing across the Pacific toward the Central and Northern Plains lost their moisture over the Rocky Mountains. Those winds reaching the prairie farms were dry, hot and filled with tons of fertile topsoil blowing east.

We learn from the past. History tells us droughts will always play a profound role in our future, and to conserve better our precious topsoil and manage our water resources. Unfortunately, history cannot tell us when the next '30s-like drought will strike.

A Blizzard by Any Other Name

Blizzards have been a part of the Dakota Territory for thousands of years, giving the Dakotas the nickname "Blizzard States." No one knows the origin of the term "blizzard," but you can bet it started somewhere up here in the Northern Plains.

We do know the root of the word blizzard stems from the act of violence. An early Webster definition describes blizzard as "a violent blow or a volley of shots." Davey Crockett once professed, "I saw two bucks and took a blizzard at one of them and up he tumbled."

I believe the phrase "volley of shots" depicts a blizzard quite well. Imagine blizzard-strength winds projecting small ice pellets onto your exposed skin at about 70 miles per hour to 80 miles per hour. In my mind, that pretty much exemplifies "a violent volley of shots."

Webster's more recent dictionary defines blizzard as "a dry, intensely cold violent storm with high winds and driving snow." The part about "a dry storm" holds for most blizzards, but not all. It's not uncommon that some blizzards produce sizeable moisture-laden snowflakes that pile up to four- to five-foot drifts in many places. During the snowstorm Atlas, on October 3 through October 5, 2013, the water weight of the snow combined with gale-force winds snapped in half tens of thousands of trees like pretzel sticks.

The National Weather Service defines a blizzard when wind speeds reach or exceed 35 miles per hour and visibility drops to less than one-quarter mile due to snow or blowing snow, and these conditions must last at least 3 hours. Snow need not be falling from the sky to have blizzard conditions.

Typically, the most catastrophic blizzards recorded in the Northern Plains develop from the rapid and violent intrusion of dry and cold air. These blizzard-making bodies of air evolve over the barren Arctic landscape where, for whatever reason, they remain fixed in place for an extended period.

If this chunk of air remains unmoved at those high latitudes, it unceasingly loses heat due to the winter's shortened days and long cloudless nights. The air mass gets intensely cold, acutely dry, and markedly dense and heavy.

So now there sits a large mass of frigid and dense air waiting for some slight change in the upper-level weather pattern to trigger a sudden southward flow of the now-frigid air. Often a slight shift in the jet stream location can cause the icy body of air to break loose and avalanche down across the Northern Plains.

More times than not, the wind pattern keeping the frigid air in place over the Arctic allows for a dry, sunny and mild weather pattern across the Northern Plains. With only a slight shift to the prevailing wind pattern, those warm and calm days quickly become bitter cold days with deadly consequences.

History reports that the most catastrophic Northern Plains blizzards caught many people and animals unprepared for the bitter cold and biting winds. The following two sub-chapters discuss two unexpected 1888 blizzards—one dry and one wet, but both killers.

The Tale of Two Killer Blizzards

Two blizzards blasted the United States in 1888, taking the lives of nearly 2,000 people, some old, and many young. Two different storms, yet the same. One storm lasted three days, carrying with it lots of heavy snow. The other continued only a half-day with little snow. The common thread tying the two storms together included cold temperatures, high winds and, most importantly, no forewarning.

The most famous blizzard that year hit the Northeast on March 12, 1888 and lasted until March 14, 1888. An intense area of low pressure moved up the east coast just offshore carrying abundant amounts of heavy snow. These "Noreaster" storm systems pumped copious amounts of Atlantic Ocean moisture across the New England region with powerful northeast winds.

The 1888 Blizzard in New York, NY (Google Images)

This 1888 blizzard paralyzed an area from Maryland to Maine, bringing trains to a standstill, silencing telegraph and telephone services and closing businesses. Major cities such as New York quickly became isolated from the rest of the country as if located on another continent. When the storm finally moved out to sea, rural folks found themselves wholly stranded for weeks, buried under snow that covered the treetops. Sadly, in New York alone, more than 800 funerals and burials were waiting for the snow to melt.

The second blizzard of 1888 occurred two months prior on January 12, 1888, when in 12 short hours, strong northwest winds swept across Dakota Territory, crippling the Plains from the Rocky Mountains to the Mississippi River to Mexico. This blizzard became known as the School Children's Storm, because it began while school was in session, changing a mild, sunny day into a nightmare of blinding cold in a matter of minutes.

Wind speeds of 60-70 miles per hour were commonplace along with temperatures suddenly dropping from 60 to 70 degrees Fahrenheit above zero to 40 degrees Fahrenheit below zero, bringing wind chill readings down to 120 degrees Fahrenheit below zero. Miles City, MT, recorded 65 degrees Fahrenheit below zero, long the U.S. record cold temperature. Visibility dropped to a few yards in a matter of minutes due to blowing snow, catching almost everyone by surprise.

The following excerpt from a South Dakota farmer living in Faulk County best conveys the horror and ferocity of this storm as he drove his team to the water well about a half-mile from his house: "Old Bawly threw up her head and snorted ... and started for the barn as fast as she could tear. I felt a puff of wind. It was cold and from the northwest, the storm quarter. A hurried glance showed me a gray bank. I straightened the team toward home directly north, took a hurried glance around to locate myself, and muttered a prayer. Then the wind came. Everything was blotted out, the trail disappeared, the horse's heads were not visible. It was difficult to breathe and utterly impossible to keep my eyes open against the driving snow. The cold was piercing ...I was already lost. To the right then, either ahead or behind, lay my stubble field which the drifting snow could not as yet have covered... If we were between the house and the well, there would be a barbed wire fence at the far end of the field. Suddenly, the team stopped, and after a brief break in the storm, I saw the most beautiful sight in the world. Two rusty wires stapled on a straggling row of posts... Good old fence... I now know why lost men would give way to panic and run until exhausted."

Fortunately, for this South Dakotan, a fence line directed him home. No such luck for hundreds of other folks, including too many school children told to leave the school early and go home, only to become hopelessly lost in the blinding snow. South Dakota reported nearly 179 deaths, and Minnesota and Nebraska reported another 500 deaths.

School Children's Storm' Ferocious, Forgotten

Let me now expand about what I think to be the most famous blizzard in the Northern Great Plains history. This most severe snowstorm, known as the "School Children's Storm," occurred on Jan. 12, 1888. I believe most folks have never heard of this storm. So, I'll use this space to bring this apparently not-so-famous storm out of the history books and, if only briefly, into our immediate conscience. Regardless of its temporary status, lasting about 12 hours, the storm took many young lives. In that short 12-hour time frame, the blizzard killed hundreds of school children across the Northern Plains.

The day the storm attacked the Plains began as a mild, comfortable day with temperatures in

The 1888 School Children's Blizzard (Google Images)

the 60s and 70s, due to a warm chinook wind sweeping across the landscape. Many children started their journey to school in light clothing appropriate for the morning's warm temperatures. Surely snow, wind and bone-chilling cold never crossed the minds or was spoken of, for that matter, among the children as they headed off for school on foot, often miles from home.

On that fatal day, that mild morning changed dramatically and quickly into a hollowing, furious, bitterly cold afternoon, turning the once recognizable landscape into a white, howling abyss. The leading edge of the blizzard moved through South Dakota around noontime and into Nebraska around 2 PM and eventually reached Texas 24 hours later where, for the first time, the Texas' Colorado River froze to a depth of 12 inches.

The front traveled southward at about 60 to 70 miles per hour, dropping temperatures nearly 100 degrees Fahrenheit in the blink of an eye. With no warning, the sky darkened, filling the air with snow and ice driven by a wind so furious that voices fell silent at six feet. Visibility quickly dropped from miles to a few feet due to blowing snow and ice.

One account of the storm's passage tells of one country school building shivering and quaking, slamming shut the window shutters from the terrific wind. In an instant, the schoolroom, like many school rooms across the Plains back then, became black as night. The sudden weather change startled many teachers into sending their children home early, which was a terrible mistake. Some teachers permitted only those children living south of the school to go home, thinking they would have little problem finding their way with the wind to their back — again a tragic blunder.

"My brothers and I could not walk through the deep snow in the road, so we took down the rows of corn stalks to keep from losing ourselves till we reached the fence. Walter was too short to wade the deep snow in the field, so Henry and I dragged him over the top. For nearly a mile, we followed the fence till we reached the corral and pens (and safely home)." (Meier)

These children were lucky to make it home. Tragically, the blinding barrage of snow, driven by strong winds, hopelessly impaired the efforts of many other youngsters to reach home safely. Battling the powerful force of the stiff, cold currents and stumbling through a white hell produced by the fierce, snowy gale, many young school-age children fell to the false sense of comfort hypothermia brings on and just curled up and froze to death.

In South Dakota, 179 deaths were reported, mostly school children, with similar numbers in Nebraska and Minnesota. No other winter storm produced such massive casualties of the frontier-children living on the Northern Plains.

Holy Week's Blizzard a Big Pain

I'll be talking about the "Holy Week Blizzard" of 2000 long after the removal of broken tree limbs, the restored electrical power, the Missouri River recedes and my sore butt recovers. On April 19 through the afternoon of April 20, between Palm Sunday and Easter, winds piled up globs of heavy wet snow almost everywhere imaginable across the Black Hills and foothills. That storm managed to squeeze out of the atmosphere one to three feet of snow, causing extensive damage, mostly snapping trees and downing power lines.

The day before started out mild with light to moderate rain showers beginning about 8 PM moving into Southwestern South Dakota. An area of low pressure that soaked California a few days earlier started having its effect on the Northern Plains. Once the low-pressure system crossed the Rocky Mountains and into the Eastern plains of Wyoming, it mushroomed into a powerful storm with snow, fierce winds and blowing snow swirling around its center.

The tremendous growth of the storm resulted from complementary atmospheric conditions at both the surface and aloft. The inward spiraling motion, converging at the surface, forced the air to rise over the center of the low. The rising currents caused the air temperature to drop in the storm's center, clouds to condense and precipitation to fall; first as rain and then as very wet and heavy snow. Recorded snowfall attributed to that blizzard totaled 18 to 33 inches in the Black Hills and eastern foothills, and 12 inches or less on the prairies just to the east.

Diverging air across the top of the storm enabled this storm to become even more formidable. Without diverging air aloft, the upward spiraling air could only accumulate above the storm and eventually crush it. The intense amplification of this storm depended mostly on the center of converging winds located beneath the center of the diverging winds aloft.

The storm's location relative to the Black Hills contributed yet another factor to the large amount of heavy snow across the region. The storm's position permitted northeasterly winds to push moisture-laden air up the eastern slopes. This topographic effect, common to Black Hill's area, provided an additional "wringing out" effect, depositing lots of moisture across the Northern, Southern and Eastern Black Hills. Liquid concentrations of the snow ranged from two to four inches of water.

The Holy Week Blizzard in Rapid City, SD (Rapid City Journal)

Official snow totals included 24 inches in Custer, Hill City and the Pactola Reservoir, 18 inches in Deadwood, Nemo and Ellsworth AFB, 14 inches in some parts of Rapid City, 12 inches in Hot Springs and Mount Rushmore, 8 inches in downtown Rapid City and 5 inches at the airport. By day's end, gale force winds drifted those "official" totals of heavy snow as high as four to five feet.

I know this from personal experience because I found myself on top of those drifts all the while precariously perched on the back of NewsCenter1's snowmobile. On that snowy day when automobile travel was impossible, Brent Gulbranson, back then NewsCenter1's assignment editor, assigned himself the job of transporting on-air types, like myself, from their homes to the studio via that very snowmobile. I quickly learned that Brent loved to go fast, while I enjoy slow and easy.

While Brent controlled the throttle, I worked feverishly to keep my butt directly over the snowmobile. That way, after Brent launched us off one of the many snow moguls we encountered, my butt would be firmly back in the saddle before the next mogul encounter.

Despite centuries of careful observation confirming Newton's findings of gravity, it only takes one execution to the contrary to disprove it. I believe if Newton rode on the back of Brent's snowmobile on that day, he too would have cause to ponder the validity of gravity's laws. After broadsiding those high moguls at the speed of Dale Earnhardt's Daytona 500-mile qualifying time, Brent and I found ourselves at, what appeared to me, right angles to the ground. I concluded that at that moment, Newton's law about gravity no longer existed. How else could Brent right the airborne snowmobile's seat in time to place it back under my butt?

The April 10-12, 2019 winter storm snowfall in Rapid City, SD

So, either Newton was wrong after all these centuries, or Brent exhibited super-human driving skills. No matter, my butt hurt for days afterward.

More recently, during the April 10 -12, 2019 winter storm, the atmospheric structure and positioning of an intense spring blizzard, replicated the Holy Week's atmospheric makeup nicely again producing copious amounts of snow across the Black Hills. Like the Holy Week's storm, this storm's moisture content and location relative to the Black Hills influenced the distribution of snowfall amounts throughout the region considerably.

Storm snowfall totals ranged from 30 inches over the Northern and Central Black Hills to a couple of inches over the Southern Black Hills. A few miles

SSE of Terry Peak collected 30 inches of the white stuff, Lead recorded two feet of snow, Johnson Siding 17 inches, Sturgis 14 inches, nearly 13 inches fell a few miles SW of Rapid City, and 7.4 inches landed on the heads of Mt Rushmore. But this time I was able to arrive at the TV station without any butt discomfort, thanks to a four-wheel-drive vehicle and careful timing.

Atlas Failed, News Prevailed

During my 20 plus years at NewsCenter1 NewsCenter1, we have never missed one newscast due to equipment failure or weather; never. The reason for our success rate rests on the shoulders of our Operations Manager, Mark Walter, whose knowledge, skill and creativity would always find a way around any potential show-stopping dilemma, thus putting us on the air and on time. I must say that Winter Storm Atlas tried its best to prevent one newscast from going on air. I'm pleased to report that Atlas failed and our 10 PM newscast, that memorable evening, went out across the airways and on time.

First a little refresher about Winter Storm Atlas, which brought the Black Hills region to a complete and utter standstill from October 3 to October 5, 2013. This one storm brought feet of snow, sustained winds of 44 mph with 70 mph gusts, downed tens of thousands of trees, caused widespread power outages for days, killed thousands of livestock and stopped dead in its tracks the day-to-day goings-on of the region.

Deadwood recorded 48 inches of snow, Sturgis and Piedmont 35 inches, southwest Rapid City 31 inches and Spearfish 26 inches of heavy wet snow. The official snowfall in Rapid City measured 23.1 inches, making Atlas the second most intense snowstorm on record in the city.

While that blizzard produced much disarray outside the walls of NewsCenter1, we found ourselves fatigued and woefully understaffed, but able enough to deliver a newscast. Here's how. The storm began Thursday evening but unleashed its full fury Friday evening, giving the entire staff a window to get to work that Friday morning and afternoon. After the Friday 10 PM newscast, getting back home became problematic to five of us, who found their cars buried in snow and the city paralyzed. The rest of the staff had people waiting with four-wheel-drive vehicles of all sizes and shapes to get them safely home. After an hour of trying to dig out of the now ever-growing snowbank, we quickly concluded that the studio would be home for the five of us that evening. Steve Bradshaw, Master Control Operator, stayed up through the night keeping NewsCenter1 on the air. The rest of us found a warm corner to place our cots for a long night's sleep. The winds blew, and snow kept falling all night.

The following Saturday morning, after staring out the window with genuine amazement across the vast snowscape, the five of us understood our fate for that day. No one can reach us, nor can we reach anyone until the city cleared the streets, sometime that evening. That morning two couples, who spent the night trapped in their car about a half block from the station, knocked on our door looking for warmth, which we gladly provided.

So, the five of us, Steve Bradshaw, Justin Wickersham, Zack Horn, Jerry (the last name escapes me) and myself, got to work to put on a newscast that would typically require 12 people. Steve, in Master Control, kept the station on the air, firing up the backup generators whenever we lost power, plus running all the newscast video segments and commercials. Zack, with limited directing experience, operated the control panel from the director's chair. Without an audio operator, all mics were left open. Justin produced, reported, and anchored the newscast. He interviewed and reported on the two stranded couples. Jerry worked the teleprompter. I did an extended weathercast, plus operated studio cameras when not doing the weathercast. That Saturday evening's 10 PM 35-minute NewsCenter1 newscast went out across the airways, without a flaw, to those lucky few Black Hills' residences still with electrical power in their homes.

Snow accumulations from the 2013 Atlas Blizzard in Rapid City, SD

Little did the home viewer know of the behind the scene struggles that went into the making of that newscast. A combination of nothing more than a wing and a prayer, flying alongside a can-do attitude, crafted that evening's newscast. Later that evening, after a few of the main roads were plowed and well after the broadcast, none other than Mark Walter once again came to our rescue in his four-wheel vehicle to take us home. At least in my case, close enough to where I could trudge through the knee-deep snow the rest of the way safely.

Flood in '72 Fit Like a Puzzle

During the 30th-anniversary memorial observance of the June 9, 1972 Rapid City flood, Jim Miller, retired meteorology professor with the South Dakota School of Mines and Technology, talked about the day's multiple weather events that came together to give birth to the flash-flood catastrophe that took place that evening. On that one night, the many minor pieces of the weather puzzle fit together, creating one of the most efficient and deadly weather events this country ever witnessed. Given the proper circumstances, any one of these pieces would have caused thunderstorms to develop, but when joined together, all hell broke loose.

On that fateful day, an abundance of water vapor saturated the air, making it feel more like a tropical day along the Gulf of Mexico coast. Dew point temperatures, which measure water content in the air, reached incredible levels for West River of 70 degrees Fahrenheit across the prairies just east of the Black Hills. The airport reported fog early that June morning. A ridge of high pressure stretched from Eastern North Dakota southward across Eastern South Dakota, with its center just north of North Dakota.

The wind circulation around this imposing mound of exceptionally moist air generated strong low-level winds sweeping across the prairies from east to west. These winds effectively and continually carried the humid air from the South Dakota prairies to the rugged and rocky eastern slopes of the Black Hills. Millions of years of heavy rainfall carved out deep and narrow canyons sloping eastward past Hill City, Keystone, Nemo, and Johnson Siding toward Rapid City, Box Elder, and Sturgis.

The atmosphere's instability — or amount of "boil" in the atmosphere — reached levels that often produce tornado-bearing thunderstorms. Like a pot of boiling water, afternoon temperatures rising to the 80s caused the air to churn mightily as the warm air near the surface raced skyward carrying copious amounts of water. A weak surface trough lay stationary across the Black Hills, offering a focal point for the soon-to-develop downpour. The fixed eastern slopes of the Black Hills directed and confined the fast-rising moist air into one area as well, a stretch of terrain that lay approximately along Highway 385 from near Hill City toward Deadwood.

Some of the destruction from the 1972 Rapid City flood (Google Images)

As bad luck would have it, even greater instability came into play early that evening as an upper-level pocket of cold air moved over the top of the Black Hills. Cold air over warm air means more buoyancy and energy to the rain-making process, which, at that time, already exceeded high levels.

Finally, the most significant piece of the puzzle fell into place far overhead. Without this final part of the puzzle, the horrific events that evening might never have happened. Strong winds aloft that typically move thunderstorms off the Black Hills and into the prairies to the east at speeds of 25 miles per hour to 35 miles per hour were absent. With only light winds aloft, these now highly efficient thunderstorms stayed locked in place from about 4 PM until around midnight on that horrific day.

Typically, heavy rain showers convert about 20 percent of the air's water vapor into liquid water, whereas severe thunderstorms can convert up to 50 percent of the water vapor into liquid water. On the night of June 9, 1972, those thunderstorms that lined up along Highway 385 converted a staggering 80 percent of that humid prairie air into liquid water. The storms dropped 12 inches to 15 inches of torrential rains directly into those deep canyons, pushing creek levels to historic heights. Among them, Rapid Creek fed into Canyon Lake, a lake held back by only an Earthen dam that was unable to restrain this unstoppable force of moving water, which moved homes and automobiles like corks and bent steel rails like paper clips.

I spent that night at a friend's house on the west side of Rapid City a few blocks east of Meadowbrook Elementary School and the obliterated neighbors just west of the school. When I left my friend's house that evening to drive home through the gap, rushing floodwaters moving through Baken Park and into the gap between, Cowboy Hill and Dinosaur Hill, prevented me from going any farther. The next morning, I witnessed total and complete destruction all along the Rapid Creek watershed. Houses and businesses were destroyed, with homes now sitting on streets, and cars stacked upon one another like playing cards.

The 1972 Rapid City flood aftermath left cars stacked on top of each other (Rapid City Journal)

The flood took the life of 238 men, women and children and caused 3,057 injuries. Many people totally unaware that around midnight that evening, while most were in bed, a wall of water would wash through their neighborhood. Homes damaged numbered about 2,820, with more than 200 businesses ruined, and around 5,000 vehicles demolished. Hopefully, this city will never forget that evening and keep commercialization at bay from the now beautiful green belt along Rapid Creek.

Individual pieces to this weather puzzle happen from time to time triggering strong thunderstorms. Only when all of these individual weather pieces come together, can we expect the reoccurrence of another event like the 1972 Rapid City flood — and it will reoccur. And when it happens again, Rapid City will be better prepared.

Nature's Clock Ticks Without the Y2K Bug

Thinking back a number of years, we welcomed the 21st century and the next 1,000 years – known as the millennium or just Y2K. For the first time in my life, the "out with the old, in with the new" cliché had a sit-up-and-take-notice significance.

Many folks feared whether or not we could conquer the infamous "Y2K bug," brought to the forefront mostly by the media. Y2K characterized the Year 2000. The "Y" stood for (Y)ear, the "2" represented the number itself, and "K" meant (K)ilo or 1,000, as in kilowatt or 1,000 watts. The Y2K bug was not the creepy-crawly variety you may occasionally find on your kitchen floor, but the imagined electronic bug lurking in those silicon chips found in all computers. The very same chips that now make up the fabric of our scientific, financial, military, medical, utility, and commercial world. People feared that embedded in those chips, a "virtual Y2K larvae" waited for the year 2000 ball to drop, signaling the Y2K bug to emerge and basically return our computer-dependent world back at least 100 years.

Despite HAL's insistence in "2001: A Space Odyssey," that computers can never be wrong, people feared that at the stroke of midnight on the eve of December 31, 1999, the Y2K bug would cause all computers to think the year 1900 started or the year 1800 and not the year 2000. When the computer's internal clock tick-tocked into the year 2000, would it successfully distinguish between the last two zeros in 2000? Imagine how confusing it would be for those great thinking machines to learn they were back in a time when artificial intelligence was not artificial — a time when the only data processor was the human brain, a pen, and paper. Oh, the horror of it all.

As it turned out, only a few computers leaped off the deep end into oblivion. Weather computers made their leap into the virtual time warp without a hiccup. As I see it, Mother Nature could care less about our humans arbitrary start of the calendar. Those atmospheric progressions that Mother Nature used to create weather for ages will be the same methods for creating the weather today and in the future.

The broad brush mathematical thermodynamic processes that a computer model uses to simulate a Black Hills snowstorm this year will be no different than the comprehensive mathematical thermodynamic processes that make up the past snowstorm in the year 1900 or the future snowstorm in the year 3000. The white stuff would form and fall from the sky, stick to the ground for a few days and eventually melt, pretty much the same way nature has been doing it for thousands of years.

Time Marches On, and Back

On occasion, the day's forecast for rain brings bright blue skies instead. That glorious sunshine predicted for picnic weather may inexplicably become monsoonal. But one meteorological prophecy is always right on the money: the sunrise and sunset times. Readily available printed tables tell us the exact time the sun will glimpse over the eastern horizon, keeping us on time for Easter's sunrise service. And, those same tables tell us when to plan a romantic sunset dinner in the Black Hills. To keep a little bit of art in forecasting sunrise and sunset times, we must know the dates when daylight saving time starts and ends.

Let's discuss the semi-annual daylight savings time (DST) ritual, better known by the mnemonic device, "fall back/spring forward." Once a year, sometime in November, we fall back, moving the clock back one hour and saying goodbye to DST until sometime in March, when we can recover that lost one hour. This time of year, we dig out those user manuals to decipher the "any-six-year-old-can-follow" set of instructions for resetting all the digital clocks found on the various recording, and playing devices, stoves, microwaves, watches, and automobiles back one hour. Only the grandfather clock at grandma's house allows me to "fall back" without the guidance of a mechanical engineer.

But now, in this modern age of digital clocks everywhere, with their own set of instructions translated from the original Mandarin Chinese, it begs the question, in my mind at least, why DST anyway? Can this country not just do away with this, fall back one hour/spring ahead one-hour, semi-annual ritual?

The official DST arrival and departure hour are 2 AM when most normal people are warmly tucked away for the night. If I play by the rules as established by the Uniform Time Act of 1966, I set my alarm for 2 AM at which time the alarm wakes me up to set the clock back one hour to 1 AM. I can then go back to sleep, thinking I have done my civic duty. But with my luck, the alarm would go off a second time at 2 AM, because I forgot to turn it off the first time. Who can think clearly or even rationally seconds after being jolted awake at 2 AM to perform some task that makes no sense anyway? Can you visualize the possibility of being stuck in an infinite time loop?

How did this madness known as daylight savings time start? Well for starts, it had nothing to do with the wants and wishes of the farming community. Farmers generally oppose daylight savings time, because typically their workday starts and ends with the sun, and humankind's fiddling with time in no way alters sunrise and sunset times. Many farmers find schedules inconvenienced by having to change all their routines to sell their crops to people who observe DST.

DST started during World War I, primarily to save fuel by reducing the need to use artificial lighting. Ever since the introduction of DST, it has been the object of constant legislative tinkering. In 1973, DST was year-round, but Congress had to repeal it due to an apparent increase in the number of school bus accidents during pre-dawn hours. When DST begins in April, traffic accidents invariably increase by about 8 percent, probably the result of so many drivers getting an hour less sleep. The current system of beginning DST at 2 AM on the first Sunday in April and ending at 2 AM on the last Sunday in October was not standard until 1986.

While South Dakota and most other states perform this ritual, Arizona, Hawaii and most of Indiana can discard their digital clock-user manuals with a clear conscience. They don't mess with DST. Now I wonder: When the rest of us changed our clocks in observance of DST, did Arizona automatically change time zones, possibly from Mountain Time to Central Time? Or did it go the other direction, finding itself in Pacific Time? Or, maybe none of the above? The answers could be spelled out in one of these manuals. Now, if only I was six years old and knew Mandarin Chinese.

Time Marches On, Deux

I am pleased to announce that a few organizations function quite smoothly without performing this twice-yearly two-step. One good example is the National Weather Service (NWS), an organization that removed the move-the-little-hand-in-move-the-little-hand-out shuffle from their dance card long ago.

The Meteorologist In Charge (MIC) at the Rapid City NWS, can call any other NWS office anywhere in the U.S. (or, in the world, for that matter), and set up a telephone conference for a given time, without having to calculate the difference by time zones and the give and take of daylight savings time.

The MIC might say something like, "Let's get together at 19:00 GMT for a telephone conference." And, whether he's talking to New York, Idaho, Alaska or Arizona NWS offices, everyone involved will know when to expect the telephone conference. Everyone will be online and ready to go at the same moment, i.e., 19:00 GMT.

Not so simple for the rest of us intra-or-international telephone users. We must first clarify the time zone in the other party's location, whether Pacific, Mountain, Central, or Eastern. Then, we must determine if daylight savings time becomes part of the conference schedule. Unless, of course, we're talking to Arizona, or some areas of Indiana, or Hawaii, where there is no savings time. Then we must decide whether to spring forward or fall back through the time savings hoop. Finally, to sum up, someone needs to sit down at a computer to calculate the correct time each conferee needs to be online, based on time of year, their time zone and their state, and for all I know, their astrological sign. Okay, so I exaggerate just a little.

As I stated earlier, the NWS across the country uses GMT, the Greenwich Mean Time. Military and aviation buffs call it ZULU time. By whichever, GMT/ZULU has been the standard time worldwide since 1884. The local time in Greenwich, England, establishes the Greenwich Mean Time. That time does not change back and forth during the year. It remains the figurative clock equivalent of the Rock of Gibraltar. The time 1900 GMT in Rapid City, which may be 1 PM Mountain Standard Time on the local clock, is the same as 1900 GMT in Paris, France, which may be 8 PM Local Standard Time on the French clock.

Technically, there are 24 world time zones, plus the International Dateline time zone, each numbered from minus 12 through 0 (Prime Meridian) to plus 12. Each time zone represents 1 degree of longitude as measured east and west from the Prime Meridian of the World at Greenwich, England. Most of these 24 time zones use a three-letter abbreviation (e.g., MST or Mountain Standard Time) for identification. The NWS and the military use the one designation, i.e., GMT.

So, when the MIC requests a conference call, say 19:00 GMT, both the calling and receiving parties know precisely the time of the call, even though their clocks on the wall may display different times. When you consider the astronomical amount of data collected by the NWS offices all over the world, the wisdom of dealing in only one standard time becomes apparent. Such a simple concept that works so well.

Weather an Inauguration Wild Card

George W. Bush's inauguration ceremony came and went pretty much without difficulty. With the oath administered, speeches given, hands grasped, bands played, and we welcomed our 43rd president. The only misfortune to this otherwise auspicious occasion had to be the weather.

The National Weather Service's official observations showed that while George W. had his right hand raised, the temperature was about 36 degrees Fahrenheit with 96 percent relative humidity, cloudy skies, and a brisk north wind. Add to the mix the reports of rain, snow, and fog during the inauguration proceedings, even becoming president can take second place to a warm fire, a good book and a relaxing beverage of your choice.

When compared to "normal" inauguration weather, the weather on George W's day was indeed frightful. Typically, the weather for a January inauguration in Washington D. C. would include partly cloudy skies, a 10 miles per hour wind and less than a 20 percent probability for any precipitation. Even if it did rain or snow, it would more than likely do nothing more than wet the pavement.

Even the four presidents carved into Mount Rushmore experienced better weather days during their own inaugurations. Clear, cool, and dry conditions reigned during George Washington's ceremony in New York City on April 30, 1789. During his second inauguration on March 4, 1793, in Philadelphia, hazy sunshine, light winds, and an estimated noon temperature of 61 degrees Fahrenheit greeted his ceremony.

Mount Rushmore, SD

Thomas Jefferson enjoyed mild and beautiful weather during his March 4, 1801, ceremony. The noon temperature was estimated to be 55 degrees Fahrenheit. Again, on March 4, 1805, Thomas Jefferson's celebration experienced fair weather with an estimated noon temperature around 50 degrees Fahrenheit.

In 1861, Abraham Lincoln watched the rain during the morning hours of his inauguration; however, during the afternoon, it quickly changed to sunny and mild conditions. Again, in 1865, it rained for two days right up to the ceremony, at which time it ended, and the sun broke through. The noontime temperature was about 45 degrees Fahrenheit. Total rainfall for the day was 0.30 inches, with the bulk of it falling about daybreak.

And finally, in 1905, Theodore Roosevelt enjoyed 45 degrees Fahrenheit and sunny skies with strong northwest winds. Patches of snow remained on the ground from light snow that fell the day before.

A healthy dose of hostile atmospheric reality into the presidential-installation ceremony comes as no surprise. As bad as the weather was for George W's inauguration, history tells us it just may have been much worse. President William Henry Harrison stood outside about two hours on a cloudy, cold and blustery day without a coat and hat. He died from pneumonia within a month.

First Lady, Abigail Fillmore, wife of the outgoing President Millard Fillmore, caught a cold as she sat on the cold, wet, exposed platform during the swearing-in ceremony of Franklin Pierce. The cold developed into pneumonia, and she, too, died at the end of the month.

President William H. Taft's 1909 ceremony was forced indoors due to a storm that dropped 10 inches of snow over the capital city, bringing all activity to a standstill. Despite the freezing temperatures, howling wind, snow, and sleet, a large crowd gathered in front of the Capitol to view the inauguration. Unfortunately, they could not, as the weather forced the ceremony indoors.

Two hundred thousand visitors came to Washington D. C., for the 1937 inauguration of President Franklin Delano Roosevelt. The day was terribly cold with sleet and freezing rain falling over the crowd during the 12:23 PM ceremony. Despite the wet, the president rode back to the White House in an open car with a half-inch of water on the floor. Total rainfall for that day was a wet 1.77 inches, which still stands as the record rainfall for January 20th.

And finally, during President Ronald Reagan's January 21, 1985, second swearing-in ceremony, the bitter cold temperatures forced the proceedings indoors. The outside temperature at noon was only 7 degrees Fahrenheit, with wind-chill temperatures during the afternoon in the minus 10 to minus 20-degree Fahrenheit range — the coldest temperatures reported for any inauguration. The reasonable heads in charge of the inauguration canceled the parade as well.

Winter's Unexpected Beastly Savagery

A few years back a close friend of mine, Joe, his son, Tony, and I drove to Lubbock, TX, to attend Nebraska — Texas Tech football game. A trip with friends is always more about the journey than the destination.

We know that as time passes, the bad moments fade from one's memory, leaving only the often-embellished good memories. Joe and I reminisced about the great times growing up together on the streets of Chicago, IL. We did not discuss the bad times, because the mind finds a way to bury them deep in our subconscience. I've already purged the one-sided football score (in Nebraska's favor), rough roads and, of course, the high price of gas from my memory.

This selective purging process certainly applies to weather. Unpleasant wintry weather events can quickly fade, especially considering a string of mild winters. So, when winter again approaches after a few years of previously mild winters, please snap out of this gleeful bliss of expecting yet another mild winter. What better way to do that than by looking back at some of those harsh winters from years past? The High Plains weather history has documented some brutal winters, reminding us that most assuredly, more unfavorable winter weather will always find its way to our doorstep.

More than 100 years ago, one winter season struck the Northern High Plains with such cruelty and persistence that it all but annihilated the most magnificent herd of range cattle in the world. This winter calamity started on November 13, 1886 and persevered for nearly four straight months. That year the hot and dry summer wiped away any forethought of the pending savagery that the following winter brought to the frontier.

Somehow the wild critters expected the weather would soon turn dreadful. The story goes that the beavers worked long hours cutting and storing willow brush to last several winter seasons; muskrats grew unusually long winter coats and built extra tall reed houses. The geese, ducks, and songbirds headed south weeks earlier than usual.

Snow fell nearly continuously through November. A short warm spell in mid-December lasted only long enough to melt about six inches of snow, leaving a top layer of solid ice when subzero temperatures quickly returned. Cows hungry for the buried frozen grass cut their muzzles as they tried to gnaw through the ice. Many cows broke through the ice into the soft snow below only to bleed to death from lacerated legs.

By Christmas, temperatures fell to 35 degrees Fahrenheit below zero and on New Year's Day temperatures reached 41 degrees Fahrenheit below zero. Four to five feet of snow covered some parts of the South Dakota prairie. The relentless northwest winds pushed the freshly fallen snow into the canyons of the Badlands where drifts to over 100 feet transformed the once broken landscape into a flat plain.

If this wasn't bad enough, a blizzard struck on January 28, 1887. "For seventy-two hours," wrote one survivor, "it seemed as if all the world's ice from time's beginnings had come on a wind which howled and screamed with the fury of demons."

In total despair, men, women, and children located on isolated ranches either froze to death, committed suicide, or simply went mad. Thousands upon thousands of cattle froze standing up like so many statues or were just buried alive. The stronger cows were found waiting by the train tracks for discarded garbage, or they invaded nearby communities desperately thrusting their heads through windows looking for food.

This horrific winter ended on March 2, when a warm chinook wind blew in from the west. Melting snow that started as a quiet trickle quickly became a roar of swiftly moving water by mid-month. The air filled with the loud rumble of flooding water down the slopes through the cracks and into the converging creek beds to the rivers.

The rivers roared for days, choked with the carcasses of dead animals. "Countless carcasses of cattle (were) going down with the ice, rolling over and over as they went, carcasses continuously seemed to be going down while the others kept bobbing up at one point or another to replace them."

Tens of thousands of cattle were merely flushed away that winter. So ended the most savage winter of the Northern Plains, taking with it one of the most extensive range herds in the world.

So, when those thoughts pop in your head that this winter cannot get any worse. The simple answer: yes, it can.

I'm Sorry After All These Years

To the nameless farmer who works a farm near Bowdle, SD; Larry and I apologize. It was not our intent to cause you to huddle like a scared rabbit in the cab of your tractor, gazing skyward for the next onslaught. To this day, I remember that horrified look on your face, as if hellish demons possessed you when you finally vaulted from your tractor.

Your mad dash across that alfalfa field would have made the U.S. Olympic sprint team proud. Even wearing those bib overalls, the speed you exhibited sprinting between your tractor and your pickup could easily carry our Olympic relay team to a gold medal win.

And, oh yes, I believe you broke the world's land speed record for 1966 Ford pickups. The way that old truck flew down that gravel road was awe-inspiring. The gravel and dust kicked up behind that old Ford gave it the look of Craig Breedlove's Spirit of America running low and fast across the Bonneville Salt Flats.

For all that anguish, Larry and I apologize for our slight miscalculation. After all these years it's time for me to clear the slate clean.

You see, on that 1972 warm and sunny summer day, Larry and I worked for a cloud seeding project stationed out of Bowdle, SD. At the time Bowdle's population included 421 citizens plus Larry, Bob and me. Larry and Bob piloted the cloud seeding project's two twin-engine Piper Comanche aircraft, and I performed the meteorological duties and operated the project's weather radar. I believe the Bowdle airport with its two hangers and grass landing strip is located next to your farm.

On that fateful day, while flying back from Aberdeen, Larry, at the controls, and I, in the copilot seat. approached the Bowdle airport. As we approached closer from the air, we spotted you mowing the airstrip. Please understand the purpose of buzzing you with our aircraft from out of the sun was for you to drive the tractor off the landing strip so Larry could land the plane safely. Unfortunately, instead of seeing you remove your tractor from the landing strip, we detected shock when Larry buzzed you a second time to emphasize our desire to land our aircraft. We buzzed you close enough for me to see the bewilderment on your face as you peeked out the tractor's cabin door.

Larry, the pilot, became more than a little irritated when you left the tractor on the landing strip to run to your parked pickup truck for a quick getaway. Please understand, it wasn't me shaking that fist at you out the cockpit window as we flew low and fast in front of your truck as you fled down that gravel road toward the highway. I too, sitting in the copilot seat, feared for my life.

Well sir, despite my pleas to Larry for calm and composure, he landed that aircraft with total disregard of your tractor sitting about two thirds down the landing strip. As it turned out, we only had to make one bounce off the ground to get that plane over the pile of alfalfa you left stacked on the landing strip. I know you are happy to know that Larry managed to get the plane firmly on the ground in time to steer it around your tractor.

Please, understand our embarrassment when we learned that in fact, we landed in the middle of your alfalfa field located next to the Bowdle grass landing strip. From the air that precise rectangle cut of alfalfa sure looked like our landing strip.

So, let's let bygones be bygones. I will sleep better now, except for the occasional nightmare of Larry's voice screaming those two words you don't ever want to hear from a pilot just before landing — "Hang on!"

🍃 EPILOGUE

In the Prologue, I stated that weather so very often has had and continues to have a prominent role in molding and locking life's events into memory, and that weather not only acts as a backdrop but often actually shapes life's most significant life-changing encounters. Let me relay one memory that in some small way, shaped my life.

Lightning over the Grand Canyon, AZ

Weather tantalizes all our senses each day, making it so engaging to so many people. The sight of lightning jumping from cloud to cloud, or from cloud to ground, leaves us in awe. We jump at the unexpected clap of thunder ringing in our ears. We stick our tongues out to taste snowflakes. And who doesn't love the smell of the air refreshed after a spring rain? I am sure you can personalize this list with your own sensual weather examples. The one sense not mentioned above is my preferred way to experience the weather — through touch.

Thanks to a long-time dear friend of mine, I cherish weather's touch. Don Larner and I have shared each other's lives for more than 30 years. Together, we've canoed rivers, hit golf balls, attended football games, waded for bass in our BVDs (the story of another time), enjoyed Mexican food and, most importantly, talked about life over a cold pitcher of beer. Don would always remind me that there are just two kinds of beer – cold beer and warm beer.

You see, Don is one of those rare people who has a knack of giving advice without directly giving advice. He has guided me through both the tough times' life often throws your way, as well as the good times. I've learned so much from him, and just maybe he learned a little from me.

Don was well into his 90's and still raising about 30 head of cattle on his ranch near Dublin, TX when he passed. But this weather memory of mine is not about Don; it's about his life-long partner and dear wife, Margaret Larner. It was Margaret who showed me how to experience the weather through touch, and in doing so, how to live life.

Margaret was an elementary school music teacher who loved kids, music, and life. After she retired, Margaret traveled the world learning all she could about the history of music. When at home, and with Don in tow, she attended the symphonies and operas and sang in the Austin Symphony Choral. In later years, Margaret lost her sight, became confined to a wheelchair, and needed 24-hour care in a nursing home. During one of my visits with Margaret, I asked if she would like to go outside for some fresh air. She agreed.

It was a beautiful October day in Austin, TX, with a deep blue sky, a warm sun, and a cool light breeze out of the northwest. I positioned Margaret's wheelchair toward the west facing the setting sun to warm her face. I remember thinking how sorry I felt for Margaret. After all, her life had turned from living it to the fullest to living in darkness, unable to get around on her own and no longer living at home with her beloved husband, Don.

What a fool I was to think such thoughts because at that moment, Margaret smiled, and with the sun lighting her face and the cool breeze blowing through her hair, she said, "Bob, isn't life wonderful?" Now, whenever the sun and cool breeze touch my face, I'm reminded that no matter what life throws my way, it is indeed wonderful.

Cherish each day and the weather that day brings.

REFERENCES

Dearborn, D. (1998). Sunspots. The Exploratorium, 3 of 4.

Meier, O. W. (n.d.). Blizzard Of 1888. Library Of Congress, 2 of 3.

Saffer, D. a. (n.d.). Water: Science and Society - Thermal Expansion and Density. InTeGrate - Penn State University.

Verma, P. C. (2015). DETERMINATION OF CONCENTRATION OF SOME HEAVY METALS IN ROADSIDE DUST IN DAMATURU METROPOLIS WHICH CAUSES ENVIRONMENTAL POLLUTION. International Journal of Advances in Science Engineering and Technology.

GLOSSARY

A

Advection: The transfer of an atmospheric property (such as temperature or moisture) by the wind.

Air mass: A widespread body of air that has similar horizontal temperature and moisture characteristics.

Air pollutants: High concentrations of solids, liquids or gaseous substances found in the atmosphere that threaten the health of people and animals and can cause harm to vegetation and structures.

Air quality index: An index that provides air pollution concentrations.

Albedo: The percent of radiation reflected from a body compared to that which strikes it.

Altocumulus: A middle-level cloud, white or grey in color, that occurs in wave-like layers or patches.

Altostratus: A middle-level cloud, grey or bluish in color, that forms in a sheet or layer appearance. Often the sun or moon may appear dimly visible through the cloud.

Anemometer: An instrument that measures wind speed.

Anticyclone: A weather system that has a Northern Hemisphere clockwise circulation pattern. Also known as a high-pressure system.

Atmosphere: The envelop of gases surrounding Earth and held to the planet by the Earth's gravitational influence.

Atmospheric greenhouse effect: The warming of the atmosphere by its absorbing and releasing of longwave radiation while letting shortwave radiation pass through it.

Atmospheric models: Computer-generated simulation of atmospheric behavior through mathematical equations.

Atmospheric pressure: The pressure exerted by the atmosphere as a result of gravitational attraction on a column of air directly over the point in question.

Autumnal equinox: The equinox when the sun approaches the Southern Hemisphere and its noon rays are directly overhead at the Equator; the official start of autumn. Occurs around September 23.

B

Barometer: An instrument for measuring atmospheric pressure.

Bermuda high: A semipermanent subtropical high-pressure cell in the North Atlantic, whose circulation is responsible for the warm and humid conditions that prevail across the North American's Southeastern regions.

Blizzard: A severe weather condition characterized by cold temperatures and strong winds (greater than 35 mph) bearing a large amount of snow either falling or blowing.

Blowing snow: Snow lifted from the ground by the wind to a height of 6 feet or higher reducing horizontal visibility at or above that level.

Breeze: A light wind with speeds generally 4 to 31 mph.

C

Carbon dioxide (CO_2): A colorless and orderless gas found in the Earth's atmosphere that is a selective absorber of the sun's infrared radiation.

Carbon monoxide (CO): A colorless and orderless toxic gas that forms due to the incomplete combustion of fossil fuels.

Ceiling: The base-height of the lowest layer of clouds.

Celsius: A temperature scale whose ice point is 0 degrees and boiling point is 100 degrees.

Chaos: The property of a system that exhibits initial erratic behavior in small stages that can lead to large and often unpredictable behavior changes in the future.

Cirrocumulus: A type of high-cloud that appears as a thin white patch of clouds without shadows.

Chinook wind: A warm and dry descending wind found on the leeward side of a mountain range (the Black Hills for example).

Cirrostratus: A type of high-cloud that appears as a whitish veil that at times produces a halo around the sun or moon.

Cirrus: A type of high-cloud made up of ice crystals forming thin, white featherlike clouds in patches, filaments or narrow bands.

Climate: The accumulation of daily and seasonal weather events over a long period of time.

Cloud: A visible collection of tiny water droplets or ice crystals in the atmosphere above the Earth's surface.

Cloud base: The lowest level in the atmosphere where the air contains a detectable quantity of cloud.

Cloud seeding: The introduction of artificial substances into a cloud to modify its development to decrease hail amounts or increase precipitation.

Coalescence: The merging of many cloud droplets into a single larger drop.

Cold front: The leading edge of a moving cold air mass that replaces a warmer air mass.

Condensation: The process by which water vapor becomes a liquid.

Condensation level: The level above the Earth's surface marking the base of a cumuliform cloud.

Condensation nuclei: Tiny particles in the atmosphere upon which condensation of water vapor begins.

Continental air: A type of air whose characteristics are developed over a large land area resulting in low moisture content.

Convection: Motion within the atmosphere that transports or mixes certain properties of the atmosphere (e.g. moisture or energy).

Coriolis force: An apparent force observed on any free-moving object in a rotating system, such as the rotation of the Earth on its axis.

Cumulonimbus: An exceptionally dense and vertically developed cloud, often with an anvil shaped top, often producing lightning and heavy rain.

Cumulus: A type of low-level cloud that grows vertically as rising, mounds, domes, or towers.

Cut-off low: A cold upper-level low that has become detached from the basic upper-level westerly flow in the mid-latitudes.

Cyclone: An area of low pressure around which the winds blow counter-clockwise in the Northern Hemisphere.

D

Density: The ratio of a mass of any substance to the volume occupied by it.

Deposition: A process that occurs in subfreezing air when water vapor changes directly to ice without becoming a liquid first.

Depression: An area of low pressure.

Dew: Water vapor that has condensed on grass or other objects near the ground when near-surface temperatures have fallen below the dew point temperature.

Dew point temperature: The temperature at which the air must cool in order for it to become saturated.

Dog days: A period of greatest heat in summer, usually from mid-July to the end of August.

Doldrums: The region near the equator characterized by high pressure and light, shifting winds.

Doppler radar: A radar that determines the velocity of precipitation either moving toward or away from the radar unit.

Downburst: The strong downward-flowing current of air associated with a severe thunderstorm.

Drifting snow: Snow raised by the wind from the Earth's surface to a height of less than 6 feet.

Drizzle: Very small, numerous and uniformly dispersed water droplets that may appear to float and fall to the ground.

Drought: A period of abnormally dry weather of sufficient length to cause crop damage, water-supply shortages, and other serious water imbalances.

Dry adiabatic rate: The rate of change of temperature in a rising or descending dry air parcel.

Dry snow: Powdery snow from which a snowball cannot easily be made.

Dust: Solid materials suspended in the atmosphere that give a tannish or greyish hue to distant objects.

Dust bowl: A region of the United States affected by extreme drought and dust storms in the decade of the 1930s.

Dust devil (or whirlwind): A small but rapidly rotating wind made visible by the dust, sand or debris it picks up from the ground.

E

Easterly wave: A migratory disturbance embedded within the broad easterly current that moves from east to west across the tropics.

Echo: The appearance on a radar screen of radio energy reflected back from a radar target i.e. a thunderstorm.

El Niño: An extensive ocean warming that extends along the coast of Peru and Ecuador, and westward over the tropical Pacific Ocean.

Energy: The property of a system that enables it to do work.

Entrainment: The mixing of environmental air into a preexisting column of air or cloud.

Equinox: The moment at which the sun passes directly over the Earth's equator.

Evaporation: The process by which a liquid is transformed into a gas; the opposite of condensation.

Exosphere: The outer most portion of the Earth's atmosphere.

Eye: The roughly circular region of comparatively light winds and fair skies found at the center of a tropical storm or hurricane.

F

Fahrenheit scale: A temperature scale whose ice point is 32 degrees and boiling point is 212 degrees.

Fata Morgana: A process whereby an initial change in an atmospheric process will reinforce positive or negative feedback.

Flash flood: A flood that rises and falls quit rapidly with little warning or no advance warning, as a result of intense rainfall.

Flood: The condition of water overflowing the natural or artificial confines of a stream or other body of water.

Fog: A cloud with its base at the Earth's surface.

Freeze: The condition when the air temperature remains below freezing over a widespread area, often causing crop damage.

Freezing rain and freezing drizzle: Rain or drizzle that falls in liquid form the freezes when striking a cold object or the ground.

Front: The interface or transition zone between two air masses having different densities.

Frost: A covering of interlocking ice crystals on exposed surfaces when the temperature of the surface and those exposed objects falls below freezing.

Frostbite: The partial freezing of exposed parts of the body, causing injury to the skin and/or deeper tissue.

Funnel cloud: A tornado whose circulation has not reached the ground.

G

Graupel: Tiny ice particles that form in a cloud when smaller ice crystals or snowflakes collide with supper cooled water droplets.

Greenhouse effect: See Atmospheric greenhouse effect.

Gulf Stream: A warm, swift and narrow ocean current flowing along the east coast of the United States.

Gust: A sudden brief increase in wind speed, whose duration is often less than 20 seconds.

Gust front: A boundary that separates a cold thunderstorm downdraft from warm, humid surface air.

H

Hailstones: Transparent or partially opaque lumps of ice that range in size from pea size to softball size, always produced by convective clouds, such as thunderstorms.

Haze: Fine dry or wet dust or salt particles dispersed through a portion of the atmosphere that diminishes horizontal visibility.

Heat: A form of energy transferred between systems due to a difference in temperature.

Heatstroke: The physical condition induced by a person's overexposure to high temperatures.

Heat wave: An extended period of abnormally and uncomfortable hot and humid weather.

High: See Anticyclone.

Horse latitudes: The belt of latitudes at about 30 degrees where the winds are light, and the weather is hot and dry.

Humidity: A general term that refers to the air's water vapor content.

Hurricane: A tropical cyclone having sustained winds of 74 mph or greater.

Hydrologic cycle: The composite picture of the exchange of water between the Earth, the atmosphere and the oceans.

Hygrometer: The instrument that measures the humidity, or water vapor content of the air.

I

Ice nuclei: Particles that act as nuclei for the formation of ice crystals in the atmosphere.

Ice pellets (sleet): A type of precipitation consisting of translucent ice fragments that bounce when hitting a hard ground.

Infrared radiation: Electromagnetic radiation having a wavelength bounded on its lower limit by visible radiation and its upper limit bounded by microwave radiation.

Insolation: Solar radiation received at the Earth's surface.

Instability: An atmospheric condition in which certain disturbances, when introduced into a stable environment, increase in magnitude.

Instrument shelter: A boxlike structure designed to protect weather instruments from direct sunlight and precipitation.

Intertropical convergence zone: The boundary zone separating the northeast trade winds of the Northern Hemisphere from the southeast trade winds of the Southern Hemisphere.

Inversion: An increase in air temperature with height.

Ionosphere: The layer in the upper atmosphere where high concentrations of ions and free electrons exist.

Isobar: A line connecting points of equal pressure.

Isotherm: A line connecting points of equal temperature.

J

Jet stream: Relatively strong winds concentrated in a narrow band in the upper atmosphere.

K

Kevin scale: A temperature scale with zero degrees equal to the temperature at which all molecular motion ceases.

Knot: A unit of speed equal to 1 nautical mile or 1.15 mph.

L

Lake-effect-snows: Localized snowstorms that form on the downwind side of a lake.

Lapse rate: The decrease of temperature with height.

Latent heat: The heat that is either released or absorbed by water when it undergoes a change of state, such as during evaporation, condensation, or sublimation.

Lenticular cloud: A cloud in the shape of a lens.

Lightning: Any and all forms of visible electrical discharge from a thunderstorm.

Long-range forecast: Generally used to describe a weather forecast beyond about 8.5 days into the future.

Long wave: A wave of pressure in the upper level of the westerly winds characterized by long length and significant amplitude.

Low: An expression for an area of "low" pressure, referring to an area of minimum atmospheric pressure.

M

Macroburst: A strong, large area downdraft that can occur beneath thunderstorms. A smaller area downdraft is called a microburst.

Maritime air: A type of air whose characteristics are developed from large bodies of water.

Maritime polar air mass: An air mass characterized by low temperatures and high humidity.

Maritime tropical air mass: An air mass characterized by high temperatures and high humidity.

Mean temperature: The average temperature at any given location for a decade-moving-average over a 30-year period (the same for mean rainfall and snowfall).

Mesosphere: The atmospheric layer between the stratosphere and the thermosphere. Ranging in altitude between 12 miles and 50 miles.

Meteorology: The science the deals with the phenomena of the atmosphere.

Microburst: See Macroburst.

Mirage: A refraction phenomena that makes an object appear to be displaced from its true position.

Mixing layer: The unstable layer of the atmosphere that extends from the surface up to the base of the inversion.

Mixing ratio: The ratio of the mass of water vapor in a given volume of air to the mass of dry air.

Moist adiabatic rate: The rate of change of temperature in a rising or descending saturated air parcel.

Moisture: The general term referring to the water vapor content in a given volume of air.

Molecule: A collection of atoms held together by chemical forces.

N

Nautical mile: The distance unit in the nautical system, its value is 1.1508 statute mile.

Nimbostratus: A type of middle-level cloud diffuse by falling rain, snow or sleet.

Nitric oxide (NO): A colorless gas produced by natural-bacterial action in the soil and by combustion processes at high temperatures.

Nitrogen (N_2): A colorless and orderless gas that occupies about 78 percent of dry air in the lower atmosphere.

Noctilucent clouds: Wavy, thin bluish-white clouds that are best seen at twilight in polar latitudes at high altitudes.

Normal: The average value of a meteorological element over a fixed period of time.

Northern lights: A glowing light display in the nighttime sky caused by excited gases in the upper atmosphere giving off light.

Nowcasting: Short-term weather forecasts varying from minutes to a few hours.

Nucleus: A particle of any nature (salt, sand, dust, etc.) upon which molecules of water or ice accumulate as a result of phase change to a more condensed state.

Numerical weather prediction: Forecasting the weather based on solutions of mathematical equations by high-speed computers.

O

Offshore wind: A breeze that blows from the land to the water. Opposite of an onshore wind.

Onshore wind: A breeze that blows from the water out over the land.

Orographic uplift: The lifting of air over a topographic barrier, i.e. the Black Hills.

Overcast: A sky condition in which the sky cover is solid attributed to clouds or some other obscuring phenomenon.

Overruning: The condition that occurs when air moves up and over another layer of air.

Oxygen (O_2): A colorless and orderless gas the occupies about 21 percent of dry air in the lower atmosphere.

Ozone (O_3): An almost colorless gaseous form of oxygen with a weak order, found primarily in the stratosphere and in polluted air near the Earth's surface. It's the main ingredient of photochemical smog.

P

Parcel of air: An imaginary small body of air that is used to explain the behavior of air.

Particulate matter: Solid particles or liquid droplets that are small enough to remain suspended in the air.

Partly cloudy: A sky condition typified by an average cloudiness from 4 percent to 7 percent coverage typically over a 24-hour period.

Peak gust: The highest instantaneous wind speed recorded at a weather station usually over a 24-hour period.

Persistence: The tendency for the occurrence of a weather event to be more probable at a given time, when the same event occurred in the immediately preceding time period.

Polar air: The type of air whose traits are developed over high latitudes, especially within the subpolar regions.

Pollutants: Any gaseous, chemical, or organic matter that contaminates the atmosphere, soil, or water.

Precipitation: Any and all forms of water particles, whether liquid or solid, that fall from the atmosphere and reach the Earth's surface.

Pressure: See atmosphere pressure.

Pressure gradient: The rate of decrease of atmospheric pressure per unit of vertical or horizontal distance.

Pressure tendency: The rate of change of atmospheric pressure within a specified period of time.

Prevailing wind: The wind direction most frequently observed during a given time period.

Probability forecast: A forecast of the probability of occurrence of one or more mutually exclusive set of weather conditions.

Psychrometer: An instrument used to measure the water vapor content of the air.

R

Radar: An electronic instrument used to detect objects, such as falling precipitation, by their ability to reflect microwaves back to a receiver.

Radar echo: See echo.

Radiational cooling: The cooling of the Earth's surface and nearby air the results when the surface sustains a net loss of heat.

Radiosonde: A balloon-borne instrument that measures and transmits pressure, temperature, and humidity to a ground-based receiving station.

Rain: Precipitation in the form of liquid water drops heavier than drizzle.

Rainbow: An arch of concentric colored bands that spans a section of the sky when the sun is present and positioned at the observers back.

Rain gauge: An instrument designed to measure the amount of rain that falls during a given time period.

Rainmaking: A common term referring to all activities designed to increase, through artificial means, the amount of precipitation released from a cloud.

Rawinsonde: A method of observing upper-air weather conditions--notably wind speed and direction, temperature, pressure, and humidity--by means of a balloon-borne radiosonde.

Reflectivity: A measure of the portion of the total amount of radiation reflected by a given object.

Refraction: The bending of light as it passes from one medium to another.

Relative humidity: The ratio of the actual vapor pressure of the air to the vapor pressure of the air when saturated.

Ridge: An elongated area of high pressure.

Rime: A white and milky granular deposit of ice formed by rapid freezing of supercooled water droplets as they become in contact with an object in below-freezing air.

Riming: The growth of a precipitation particle by the collision of an ice crystal or snowflake with a supercooled liquid water droplet that freezes on contact. (Also known as accretion)

S

Saturation: A condition of the air in which any increase in the amount of water vapor will initiate a more condensed state within the air.

Saturation vapor pressure: The maximum amount of water vapor necessary to keep moist air in equilibrium with a surface of pure water or ice.

Scattering: The process by which small suspended particles diffuse a portion of the incident radiation in all directions.

Sea breeze: A coastal local wind that blows from the ocean onto the land.

Sea-level pressure: The atmospheric pressure at mean sea level.

Severe storm: Any destructive storm, often used to describe intense thunderstorms that produce heavy rain, large hail, tornadoes and/or powerful winds.

Shear: Most often used in meteorology to describe the variation in wind speed and direction with height above the Earth's surface.

Short wave: A progressive low pressure wave in the horizontal pattern of air motion within the Earth's atmosphere.

Shelf cloud: A dense, arch-shaped, ominous looking cloud that often forms along the leading edge of a thunderstorm's gust front.

Shower: Intermittent precipitation from a cumuliform cloud, usually of short duration but often heavy.

Sleet: A type of precipitation consisting of tiny transparent pellets of ice. (See ice pellets)

Smog: Air that has restricted visibility due to pollution, or pollution formed by the presence of sunlight.

Smoke: Foreign particulate matter in the atmosphere resulting from the combustion process.

Snow: A solid form of precipitation composed of ice crystals in complex hexagonal form.

Solar wind: An outflow of charged particles from the sun that escape the sun's outer atmosphere at a very high speed.

Solstice: Popularly regarded as the time at which the sun is farthest north or south.

Southern oscillation: The reversal of surface air pressure at the opposite ends of the tropical Pacific Ocean that occur during the major El Niño event.

Squall line: A line of thunderstorms that form along a cold front or out ahead of it.

Stability: An atmospheric condition in which a displaced parcel of air is subjected to a buoyant force opposite to its displacement.

Standard atmosphere: A hypothetical vertical distribution of temperature, pressure and density that is regarded as representative of the atmosphere.

Standard rain gauge: A nonrecording rain gauge with an 8-inch diameter collector funnel and a tube that amplifies rainfall by tenfold.

Stationary front: A front that is nearly stationary with winds blowing almost parallel and from opposite directions on each side of the front.

Station pressure: The actual pressure measured at the observing station.

Stepped leader: An initial discharge of electrons that proceeds intermittently toward the ground in a series of steps in a cloud-to-ground lightning strike.

Storm: Any disturbed state of the atmosphere that strongly implies destructive or otherwise unpleasant weather.

Stratocumulus: A low cloud predominately stratiformed, with low, lumpy, rounded masses often with blue sky between them.

Stratosphere: The layer of the atmosphere above the troposphere and below the mesosphere, generally characterized by an increase in temperature with height.

Stratus: A type of low-level cloud in the form of a grey layer with a rather uniform base; usually does not produce precipitation of consequence.

Sublimation: The transition of a substance from the solid phase to the vapor phase.

Subsidence: The descending motion of air in the atmosphere, usually associated with high pressure.

Subtropical high: A semipermanent high in the subtropical high-pressure belt centered near 30 degrees latitude. The Bermuda high is a subtropical high.

Subtropical jet stream: The jet stream typically found between 20 degrees and 30 degrees latitude between 39,000 feet and 46,000 feet.

Summer solstice: Approximately June 21 in the Northern Hemisphere when the sun is highest in the sky and directly over the Tropic of Cancer.

Sundog: A colored luminous spot produced by refraction of light through ice crystals that appears on either side of the sun.

Sunspots: Relatively cool areas on the sun's surface.

Supercooled water: Liquid water drops whose temperature is reduced below their normal freezing point without changing its liquid status.

Synoptic scale: The typical weather map scale that shows features such as high- and low- pressure areas and fronts over a distance spanning a continent.

T

Temperature: The degree of hotness or coldness of a substance as measured by a thermometer.

Thermometer: An instrument for measuring temperature.

Thermal: A small, rising parcel of warm air produced when the Earth's surface is heated unevenly.

Thermosphere: The atmospheric layer above the mesosphere where the temperature increases rapidly with height.

Thunder: The sound due to rapidly expanding gases along the channel of a lightning discharge.

Thunderstorm: A local storm produced by cumulonimbus clouds, accompanied by lightning and thunder, and usually heavy rain and strong gusty winds.

Tide: The periodic rise and fall of the Earth's oceans and atmosphere.

Tornado: An intense, rotating column of air that often protrudes from a cumulonimbus cloud in the shape of a funnel or a rope whose circulation is present at the ground.

Trade winds: The winds that occupy most of the tropics and blow from the subtropical highs to the equatorial low and is northeasterly in the North Atlantic.

Tropical cyclone: The general term for storms that form over the warm tropical waters.

Tropical depression: A tropical cyclone having a sustained wind speed not greater than 39 mph.

Tropical disturbance: A tropical cyclone with only a slight surface wind circulation.

Tropical storm: A tropical cyclone with a sustained wind speed of 39 mph to 73 mph.

Tropopause: The boundary between the troposphere and the stratosphere, usually marked by a sudden change in the rate at which temperature drops with height.

Troposphere: The portion of the atmosphere between the Earth's surface and the tropopause or the lowest 6 to 13 miles of the atmosphere.

Trough: An elongated area of relatively low atmospheric pressure.

Turbulence: Any irregular or disturbed flow in the atmosphere that produces gusts and eddies.

Twister: A colloquial term for tornado.

Typhoon: A hurricane that forms in the western Pacific Ocean.

U

Upper air: Generally the portion of the atmosphere about 5,000 feet above the Earth's surface.

Upslope fog: Fog formed as moist, stable air flows upward over a topographic barrier, such as the Black Hills.

Upslope precipitation: Precipitation formed as moist, stable air gradually flows upward over a topographic barrier, such as the Black Hills.

Upwelling: The rising of water (usually cold) toward the surface of a body of water.

Urban heat island: The increased air temperatures in urban areas as contrasted with cooler surrounding rural areas.

V

Vapor pressure: The pressure exerted by the water vapor molecules in a given volume of air.

Veering wind: A change in wind direction in a clockwise sense in the Northern Hemisphere.

Vernal equinox: The equinox at which the sun approaches the Northern Hemisphere and passes directly of the equator. Occurs around March 20.

Virga: Precipitation that falls from a cloud but evaporates before reaching the ground.

Visibility: The greatest distance that an observer can see and correctly identify prominent objects.

Volatile organic compounds (VOCs): A class of organic compounds that are released into the atmosphere from souces such as motor vehicles, paints, and solvents.

W

Wall cloud: An area of rotating clouds that extends beneath a severe thunderstorm and from which a funnel cloud may appear.

Warm front: A front that moves in such a way that warm air replaces cold air.

Weather: The condition of the atmosphere at any particular time and place with respect to its effects upon life and human activities.

Weather modification: An effort to alter artificially the natural phenomena of the atmosphere.

Weather warning: A forecast indicating that hazardous weather is either imminent or actually occurring within the specified forecast area.

Weather watch: A forecast indicating that atmospheric conditions are favorable for hazardous weather to occur over a particular region during a specified time period.

Wind: Air in motion relative to the Earth's surface.

Wind chill: That part of the total cooling of a body by air motion.

Wind direction: The direction from which the wind is blowing.

Wind shear: See shear.

Winter solstice: Approximately December 21 in the Northern Hemisphere when the sun is lowest in the sky and directly overhead the Tropic of Capricorn.

Z

Zonal flow: The flow of air that has a predominate west-to-east component.

🍃 INDEX

Symbols

2000 Jasper Fire, 89

2012 Oil Creek Fire, 89

2017 Legion Lake Fire, 89

A

agricultural drought, 126, 149, 150

air mass, 43, 81, 83, 135, 141, 206, 227, 231, 232

air pollution, 32, 35, 133, 226

Albedo, v, 55, 56, 226

Aldrin, Edwin "Buzz"
    astronaut, 7

Alps
    mountains, 188

Amazon rainforest, 1

ambient air, 15, 17, 20, 21, 37, 53, 54, 57, 61, 62, 63, 119

Antarctic, 62, 125

anvil top, 16, 97

Apollo 11, 7, 8

Apollo 13, xiii

Arctic air, 45, 52, 53, 76, 80, 81

Argon
    atmospheric gas, 18

Armstrong, Neil
    astronaut, 7

atmosphere, xi, xii, xv, xvi, 1, 2, 3, 5, 7, 8, 14, 15, 16, 17, 18, 19, 20, 23, 24, 27, 31, 32, 33, 34, 36, 39, 41, 42, 55, 57, 59, 69, 72, 73, 75, 76, 77, 83, 84, 85, 90, 91, 94, 100, 101, 102, 108, 114, 115, 117, 118, 119, 120, 124, 125, 129, 139, 140, 141, 142, 147, 159, 160, 166, 175, 192, 198, 210, 214, 226, 227, 228, 229, 230, 231, 232, 233, 235, 236, 237, 238

atmospheric trigger, 140

aurora, 27

autumn, xiv, 1, 52, 70, 71, 72, 161, 170, 171, 172, 173, 174, 178, 179, 226

Aviation Model
    weather model, 138

B

Badlands, iv, xiv, xv, 28, 29, 128, 221

Battle Creek Fire, 89

Bermuda High
    pressure system, iv, 38, 39, 88, 89, 108, 175

biological gases, 18, 32

Black Elk, xvi, 18, 83

Black Hills, v, x, xi, xii, xiii, xiv, xv, xvi, 16, 18, 23, 27, 28, 29, 35, 38, 39, 43, 44, 49, 52, 55, 58, 59, 64, 69, 71, 72, 73, 74, 75, 79, 80, 81, 82, 83, 88, 89, 90, 91, 93, 95, 98, 100, 102, 107, 108, 109, 110, 123, 127, 139, 142, 143, 144, 145, 149, 150, 153, 154, 156, 157, 160, 161, 169, 170, 171, 172, 173, 175, 176, 177, 178, 179, 180, 184, 185, 186, 190, 203, 210, 211, 212, 213, 214, 216, 217, 227, 233, 237, 250

black ice, 45

blizzard, 41, 72, 76, 78, 118, 157, 186, 189, 190, 206, 207, 208, 209, 210, 211, 212, 222

Bowdle
    South Dakota, 222, 223

Bradshaw, Steve
    NewsCenter1 Master Control Operator, 212, 213

Brewster, Wayne
    WWII veteran, 203

Bucknall, Maj. David
    Air Force pilot, 143

Bush, George W.
    president, 219

C

calorie, 20

Canada, 39, 44, 62, 67, 69, 70, 72, 76, 80, 89, 101, 148, 164, 179

Canis Major
    constellation, 176, 177

carbon, 18, 32, 124, 188

carbon dioxide
    atmospheric gas, 19, 227

chaos theory, 116, 117, 118, 140

Chapel In The Hills
    Stav Kirke Chapel, xiv

Chicken Little complex, 105

chinook winds, 71, 72, 81, 82, 83, 190

Civil War, 195

climate, 3, 4, 6, 8, 33, 51, 58, 59, 87, 118, 124, 125, 155, 156, 159, 169, 184, 187, 188, 189, 196, 250

climate warming, 58, 59

cloud droplets, 95, 96, 97, 106, 123, 227

cloud drops, 4, 91

cold front, 60, 61, 73, 98, 134, 135, 202, 235

Coleridge, Samuel Taylor
    poet, 25

Collins, Michael
    astronaut, 8

condensation, 21, 35, 82, 83, 95, 96, 227, 229, 231

condenses, 4, 49, 87, 95, 147

continental polar
    air mass, 43

continental tropical
    air mass, 43

convection, 18, 228

Coriolis force, 24, 25, 26, 68, 228

Cowboy Hill
    Rapid City, xiii, 48, 49, 215

Crockett, Davey
    frontiersman, 206

cumulonimbus
    cloud type, 15, 16, 85, 102, 198, 236, 237

cumulus clouds, xv, 15, 85, 90, 94, 120, 139

cylinder shaped rain gauge, 151

D

daylight savings time
    DST, 217, 218

D-Day, viii, xi, 194, 195, 199, 200, 201, 202, 203, 204

dense, 203

density, 17, 22, 98, 235

desublimation, 97

dew point, 214, 228

dew point temperature, 228

dog days, 1, 38, 89, 176

Douglas, Mr. C. K. M.
    RAF pilot, 200

downdraft, 98, 99, 106, 107, 230, 231

drought, vii, viii, 126, 147, 148, 204, 228

dry adiabatic lapse rate, 82

dry bulb thermometer, 53

dust, 5, 8, 28, 33, 34, 35, 36, 39, 67, 78, 95, 105, 106, 122, 136, 180, 204, 222, 229, 230, 232

Dust Bowl
   1930's drought, viii, 127, 204, 205

Dyenforth, Gen. R.G.
   rain maker, 196

Dylan, Bob
   song writer, 67, 73

E

Earth, iv, xvi, 1, 2, 3, 4, 5, 6, 7, 8, 9, 10, 11, 12, 13, 15, 16, 17, 18, 19, 24, 26, 27, 28, 31, 34, 42, 51, 56, 57, 68, 69, 72, 73, 85, 91, 93, 94, 95, 97, 98, 107, 114, 116, 122, 123, 124, 127, 128, 144, 147, 148, 156, 160, 168, 169, 170, 171, 177, 178, 187, 188, 189, 192, 193, 226, 227, 228, 229, 230, 233, 234, 235, 236, 237, 238

easterly trade winds, 25, 26

East River
   South Dakota, 52, 86, 185, 190, 191

Eisenhower, Gen. Dwight
   Supreme Commander, 194, 195

Ellsworth Air Force Base
   South Dakota, 73, 143

El Niño, vii, 148, 159, 160, 161, 162, 229, 235

Emergency Management Office, 37

energy, 1, 2, 3, 4, 5, 6, 18, 20, 21, 23, 50, 51, 53, 54, 56, 57, 61, 71, 75, 80, 81, 82, 84, 85, 96, 98, 101, 102, 108, 111, 123, 126, 140, 141, 147, 155, 156, 162, 163, 170, 177, 188, 192, 199, 214, 228, 229, 230

environment, 4, 6, 20, 21, 31, 42, 59, 71, 81, 82, 91, 95, 96, 114, 138, 144, 188, 231

equinox, 10, 75, 76, 169, 170, 178, 226, 238

evaporates, 54, 55, 62, 99, 238

evaporation, 21, 32, 62, 78, 99, 101, 147, 150, 231

F

Fata Morgana, 28, 29, 229

Festival in the Park
   Spearfish, South Dakota, 156, 157

Fillmore, Millard
   president, 220

Fires, iv, v, 38, 90

first law of thermodynamics, 2, 20, 82

flood, xiii, xiv, 88, 110, 111, 114, 164, 165, 188, 191, 192, 213, 214, 215, 229

flooding, 90, 111, 164, 191, 192, 222

floods, 90, 94, 106, 125, 141, 165, 191

flow dynamics, 22

fluid mechanics, 22

fog, xiii, 35, 48, 49, 64, 146, 157, 158, 186, 214, 219, 237

forecasts, ix, xi, 35, 60, 97, 114, 115, 116, 118, 128, 133, 136, 137, 138, 140, 141, 142, 145, 153, 156, 159, 162, 163, 165, 166, 167, 184, 192, 199, 200, 202, 232, 250

Fort Meade
   South Dakota, 195

freezing, 4, 21, 43, 44, 45, 52, 53, 57, 61, 62, 65, 80, 97, 123, 173, 180, 181, 183, 198, 220, 229, 234, 236

freezing rain
   precipitation, 44, 45, 123, 183, 220

A Weather Legacy 241

friction, 24, 47, 48, 63, 128

frost, v, 56, 57, 173, 181, 192, 193, 229

Frost, Robert
poet, 192

Froude number, 145

fruit moon or harvest moon
September moon, 10, 11

funnel, 101, 102, 104, 105, 123, 235, 237, 238

G

Galilei, Galileo, 6

Gap, The
Rapid City, xiii, 48, 49

gas, 2, 5, 7, 15, 16, 17, 18, 19, 20, 21, 22, 31, 32, 33, 49, 54, 55, 57, 73, 83, 95, 96, 122, 162, 182, 188, 196, 221, 227, 229, 232, 233

gases, iv, 18

gas pressure, 7

General Electric, 198, 199

glaciers, 58, 59, 125, 187, 188

Gleason, Jackie
actor, 9

Global Forecast System, 138

Goldilocks, 6, 51

Goldsmid, Johann, 6

grain moon or green corn moon
August moon, 11

grass moon or egg moon
April moon, 11

gravitational force, 12

gravity, 9, 12, 13, 24, 25, 47, 69, 70, 107, 110, 111, 211

Greenwich Mean Time
GMT, 218

Grizzly Gulch Fire, 89

Gulbranson, Brent
assignment editor, 211

Gulf of Mexico, xvi, 6, 43, 44, 76, 87, 88, 92, 100, 101, 108, 160, 175, 205, 214

Gulf Stream, 230

gust front, 230

H

Hadley, George
London laywer, 26

hail, 21, 84, 85, 86, 97, 98, 99, 106, 107, 108, 109, 123, 141, 175, 176, 227, 235

hailstones, 230

Hall, Captain Roy S.
tornado observer, 104

harvest moon or hunter's moon
November moon, 11

heat, xiv, 1, 2, 3, 4, 6, 7, 13, 18, 20, 21, 38, 39, 46, 47, 48, 51, 53, 54, 57, 58, 59, 61, 62, 63, 71, 72, 73, 78, 80, 81, 82, 85, 93, 95, 96, 97, 98, 108, 123, 124, 155, 156, 177, 178, 179, 181, 182, 188, 198, 206, 228, 231, 234, 237

hemisphere, 11, 24, 25, 27, 42, 68, 76, 169, 170, 177, 178, 179, 190, 192, 193, 226, 228, 231, 236, 237, 238

Herriot, Thomas, 6

hibernation, 181, 182

high pressure, 23, 39, 42, 43, 44, 45, 52, 67, 68, 69, 70, 81, 148, 214, 228, 234, 236

high-pressure, 24, 39, 44, 49, 67, 68, 69, 70, 73, 74, 76, 77, 88, 100, 134, 147, 148, 202, 226, 236

High-Resolution Rapid Refresh
weather model, 138

Hirsch, John
    meteorologist, 139
Holy Week Blizzard, 210
Holzman, Ben
    meteorologist, 200
Homestead Act, 34, 205
Horn, Zack
    NewsCenter1 Production Operator, 213
horse latitudes, 230
humidity, 54, 230
Hubble, Edwin
    astronomer, 13
hurricane, 92, 137, 230, 250
hurricanes, 4, 92, 93, 125, 137, 140, 199, 250
hydrodynamic, 135

I

Icarus
    Greek parable, iv, 17, 18
ice age, viii, 187
ice crystals, 5, 55, 56, 57, 93, 95, 97, 180, 198, 199, 227, 229, 230, 235, 236
ice pellets
    precipitation, 230
initial weather conditions, 137
inversions
    temperature, 33, 34
ionosphere, 231

J

Jefferson, Thomas
    president, 220
jelly sandwich law, 68
jet stream, v, 70, 75
jet streams, 101

June 9, 1972, flood, xiii

K

Kellogg, 3
Krick, Irving P.
    meteorologist, 200

L

La Niña, vii, 45, 159, 160, 161
Larner, Don
    mentor, iii, 124, 224
Larner, Margaret, 224
latent heat of condensation, 82, 96
light, iv, v, vi, 27, 36, 37, 61, 93
lightning, xv, 1, 78, 85, 93, 94, 99, 140, 141, 158, 198, 224, 228, 235, 236
Lincoln, Abraham
    president, 220
liquid, 20, 21, 22, 45, 49, 73, 83, 86, 95, 96, 97, 98, 106, 180, 183, 198, 215, 227, 228, 229, 233, 234, 236
Little Elk Fire, 89
Lombardi, Vince
    football coach, 178
low pressure, 23, 42, 69, 74, 157, 207, 210, 228, 235
low-pressure, 24, 42, 49, 67, 68, 73, 77, 134, 210
lunar eclipse, 10
lunar perigee, 11

M

magnetic field, 7
maritime polar
    air mass, 231
maritime tropical
    air mass, 43

Medium Range Forecast Model
weather model, 138

melting, 2, 3, 4, 17, 20, 21, 45, 53, 54, 59, 72, 81, 83, 86, 156, 164, 169, 181, 183, 189

Mesoscale Model 5th generation weather model, 138

mesosphere, 232

meteorologists, iii, xi, 107, 110, 134, 135, 137, 142, 145, 158, 200, 201, 204

meteors, 16, 27

micro-burst, 98

Midwest, xii, 34, 58, 76, 89, 100, 101, 121, 148, 159, 188, 190, 192, 205

Midwest plains, 34

milk moon or planting moon
May moon, 11

Miller, Jim
meteorologist, iii, 213

moon, xiii, 1, 8, 9, 10, 11, 12, 201, 226, 227

moon before Yule, or long night moon
December moon, 11

moon, hunger moon, or wolf moon
February moon, 11

Mount Rushmore, xv, 78, 79, 211, 219

mudslides, 90

Murphy, Teresa
hydrologist, 90

N

National Center for Atmospheric Research, 250

National Centers for Environmental Information, 250

National Hurricane Center, 92, 137, 250

National Oceanic and Atmospheric Administration, 122, 126, 184, 250

National Park Service, 78

National Weather Service, iii, xi, 62, 90, 98, 103, 105, 107, 108, 113, 126, 127, 145, 146, 150, 153, 155, 164, 166, 167, 176, 185, 189, 206, 218, 219, 250

nature, v, vi, viii, 3, 20, 34, 59, 77, 88, 93, 118, 125, 143, 144, 154, 172, 174, 181, 182, 192, 215, 216

Nebraska, 9, 74, 99, 104, 121, 192, 208, 209, 221

Nested Grid Model
weather model, 138

Newton, Isaac, 12, 178

nimbostratus
cloud type, 232

nitrogen
atmospheric gas 18, 232

nitrogen dioxide, 32

noctilucent clouds, 27, 232

normal precipitation, v, 86

North American Model
weather model, 138

Northern Hemisphere, 11, 24, 25, 27, 42, 68, 76, 169, 170, 177, 178, 179, 190, 192, 226, 228, 231, 236, 237, 238

Northern Plains, v, xiv, xvi, 1, 7, 35, 38, 39, 43, 44, 45, 58, 67, 69, 70, 71, 72, 73, 74, 75, 76, 77, 78, 80, 87, 88, 89, 97, 101, 121, 138, 142, 148, 156, 169, 174, 175, 179, 187, 204, 205, 206, 207, 208, 209, 210, 222

North Pole, 26, 130, 188

nuclei, 35, 227, 230

NWS
    National Weather Service, xii, 62, 63, 98, 99, 105, 106, 107, 108, 109, 110, 126, 127, 128, 145, 146, 150, 155, 164, 165, 167, 218, 219

O

oceans, vii, 87, 125, 148, 158

Olbers, Heinrich Wilhelm Matthias
    German philosopher, 13

Operation Overlord, 199, 203

order, vi, 117, 140

organic gases, 32

Orville, Dr. Harold
    meteorologist, 139

outflow boundary, 98

oxygen
    atmospheric gas, 18, 19, 31, 32, 36, 90, 233

ozone
    atmospheric gas, iv, 18, 19, 31, 32, 33, 233

P

Pacific Ocean, xvi, 24, 31, 43, 81, 87, 129, 148, 160, 229, 235, 237

pellets
    snow type, 4, 45, 97, 180, 198, 206, 230, 235

perspiration, 55, 62

Petterssen, Sverre
    meteorologist, 200, 204

Pikes Peak, 18

Post, Charles William
    entrepreneur, 196

Powers, Edward
    historian, 195, 198

pressure, xi, 3, 4, 7, 14, 15, 18, 20, 23, 24, 26, 27, 39, 40, 41, 42, 43, 44, 45, 46, 47, 48, 49, 52, 57, 58, 67, 68, 69, 70, 71, 72, 73, 74, 76, 77, 79, 81, 82, 88, 89, 91, 100, 104, 114, 117, 119, 128, 133, 134, 139, 146, 147, 148, 155, 157, 158, 175, 198, 202, 203, 207, 210, 214, 226, 228, 231, 233, 234, 235, 236, 237

Przybyslawski, Col.
    Air Force pilot, 143

R

rain, xiii, xiv, 1, 4, 6, 21, 23, 27, 34, 36, 37, 43, 44, 45, 49, 51, 60, 61, 81, 85, 86, 87, 88, 90, 91, 96, 97, 98, 99, 101, 107, 108, 113, 119, 122, 123, 126, 127, 135, 139, 140, 141, 142, 145, 146, 147, 148, 149, 150, 151, 152, 153, 154, 160, 161, 169, 175, 176, 183, 184, 195, 196, 197, 198, 199, 203, 205, 210, 215, 216, 219, 220, 224, 228, 229, 232, 234, 235, 236

raindrops, 4, 21, 37, 45, 60, 61, 91, 97, 123

rain-enhancement, 195, 197

rainmaking, 196, 197, 214

rain shower, 4, 85, 91, 96, 97, 145, 176

Rapid Creek, xiii, 22, 23, 41, 111, 149, 215

Rapid Refresh Model
    weather model, 138

Rapid Update Cycle
    weather model, 138

Rayleigh scattering, 36

Reagan, Ronald
    president, 220

remote sensing weather measuring instruments, 146

Roosevelt, Franklin Delano
    president, 220

Roosevelt, Theodore
	president, 149, 220
rose moon, flower moon, or strawberry moon
	June moon, 11

S

salt, vii, 87, 182, 188, 222
Sanders, Susan
	meteorologist, 63
sap moon, crow moon or Lenten moon
	March moon, 11
satellite, 7, 16, 19, 24, 25, 31, 79, 122, 123, 124
satellites, 25, 122, 128, 133, 153
saturation ice vapor pressure, 198
scatter light, 35, 36
Schaefer, Vincent
	meteorologist, 198
Scheiner, Christopher, 6
School Children's Storm
	1888 blizzard, viii, 207, 208
seasons, 4, 21, 52, 118, 149, 168, 169, 170, 172, 185, 186, 187, 221
second law of thermodynamics, 2, 34, 50, 51, 53, 54, 62, 77, 140, 141
selective scattering, 36
severe weather warning, 103, 106, 108, 109
severe weather watch, 106, 107, 166
Siberia, 44, 45
Siple, Paul A.
	explorer, 62
Sirius
	star, 176, 177
Skyline Drive
	Rapid City, xiii, 48, 49
sleet
	precipitation, 21, 44, 45, 86, 123, 220, 230, 232
smoke
	fire, iv, 34, 37, 235
snow, v, 55, 56, 80, 81, 114, 178, 180, 206, 213, 221, 227, 228, 235
snowfall, 44, 45, 73, 89, 157, 159, 164, 178, 179, 184, 185, 190, 191, 210, 211, 212, 232
snowflakes, 37, 55, 56, 154, 179, 180, 181, 198, 206, 224, 230
snow grains, 180
solar cycle, 177
solar eclipse, August 2017 total, 9
solar eclipse, 9, 10
solar radiation, 19, 32, 57, 59, 156, 178
solar system, 1, 9
solstice, 11, 171, 177, 178, 179, 192, 193, 236, 238
soot, 33, 34, 36, 89, 91, 188
South Dakota, v, vi, ix, x, xi, xii, xiii, xiv, xvi, 11, 16, 21, 22, 37, 39, 51, 55, 58, 59, 61, 64, 73, 74, 79, 80, 81, 86, 89, 90, 96, 101, 102, 103, 104, 106, 107, 108, 109, 118, 121, 122, 127, 135, 139, 143, 145, 147, 148, 154, 157, 160, 161, 171, 172, 175, 176, 181, 183, 191, 204, 208, 209, 210, 213, 214, 218, 221, 250
South Dakota School of Mines and Technology, ix, xii, xiii, 59, 127, 139, 213
Spearfish, xv, 71, 73, 98, 99, 104, 156, 157, 158, 172, 212
Spearfish Canyon, xv, 104, 158
Spencer
	South Dakota, vi, 102, 103
spring, vii, 16, 21, 175, 192

squall line, 235

Stagg, Capt. James Martin
    meteorologist, 200

stepped leader
    lightning, 235

Storm Prediction Center
    Norman, Oklahoma, 107, 108, 109, 110

straight-line winds, 1, 73, 94, 98, 99, 106, 109

stratosphere, 16, 232, 233, 237

streamer
    lightning, 94

summer, xiv, 1, 4, 11, 22, 27, 29, 32, 38, 39, 43, 52, 53, 54, 56, 70, 71, 72, 76, 78, 88, 89, 97, 99, 102, 106, 107, 109, 141, 145, 149, 150, 151, 159, 161, 168, 169, 170, 174, 176, 177, 182, 187, 193, 198, 221, 223, 228

sun, 1, 2, 3, 4, 5, 6, 7, 8, 9, 10, 11, 12, 13, 15, 16, 17, 18, 19, 20, 28, 31, 35, 36, 38, 48, 51, 56, 57, 58, 75, 76, 80, 85, 95, 98, 101, 108, 122, 129, 142, 147, 155, 156, 157, 168, 169, 170, 171, 173, 174, 175, 177, 178, 179, 188, 189, 192, 193, 216, 217, 220, 223, 225, 226, 227, 229, 234, 235, 236, 238

sunrise, 35, 48, 216, 217

sunset, xv, 35, 72, 75, 103, 216, 217

sunspots, 7, 225, 236

supercell, 85

supercomputers, 101, 118, 128, 136, 139, 140, 152, 158

supercooled, 45, 198, 199, 234

Swigert, Jack
    astronaut, xiii

T

Taft, William H.
    president, 220

temperature, ix, xi, 7, 8, 15, 16, 17, 18, 20, 21, 25, 32, 33, 34, 37, 38, 39, 42, 44, 45, 48, 49, 50, 51, 53, 54, 56, 57, 58, 59, 61, 62, 63, 64, 70, 71, 73, 77, 79, 80, 81, 82, 83, 113, 114, 116, 117, 119, 123, 124, 125, 126, 128, 135, 139, 142, 145, 146, 148, 155, 156, 159, 160, 161, 162, 163, 171, 172, 176, 177, 178, 180, 181, 183, 184, 185, 186, 187, 189, 190, 208, 210, 219, 220, 226, 227, 228, 229, 230, 231, 232, 234, 235, 236, 237, 250

Thanksgiving turkey, 46

thermodynamic, 49, 135, 216

thermodynamics, 2, 20, 34, 50, 51, 53, 54, 62, 77, 82, 95, 113, 137, 140, 141

thermosphere, 16, 232

thunder moon or hay moon
    July moon, 11

thunderstorm, x, xi, 4, 15, 16, 21, 73, 78, 85, 93, 94, 96, 97, 98, 99, 100, 101, 102, 106, 107, 108, 109, 121, 122, 123, 144, 228, 229, 230, 231, 235, 238

thunderstorms, 1, 4, 6, 16, 67, 79, 85, 90, 95, 98, 99, 102, 106, 108, 109, 110, 120, 122, 123, 125, 130, 138, 141, 143, 144, 151, 169, 175, 213, 214, 215, 230, 231, 235

tipping bucket rain gauge, 150

tornado, 1, 84, 85, 98, 99, 100, 101, 102, 103, 104, 105, 106, 108, 109, 116, 123, 176, 214, 230, 237

Tort Claims Act, 167

tropics, 5, 6, 26, 229, 237

troposphere, 2, 5, 6, 15, 16, 17, 18, 20, 21, 22, 27, 37, 82, 97, 103, 113, 123, 127, 128, 236, 237

trough, 237

Tyndall, John
 physicist, 37

U

ultraviolet radiation, 19, 31

universe, 1, 9, 12, 13, 19, 125

upper-level winds, 6, 75

Upper Plains, 39, 52, 76, 78, 108, 147, 186, 191

upslope, 49, 145, 157, 158, 160

upslope winds, 145, 157, 160

V

Vanocker, 172

visibility, xiii, 33, 146, 186, 201, 206, 227, 230, 235

volcanoes, 91

W

Wally
 my dog, 22, 41, 58, 77, 120, 144, 180

Walter, Mark, 212, 213

warm front, 134, 135

WARNING, 37, 38, 47, 63, 103, 105, 106, 108, 109, 111, 118, 129, 166, 172, 209, 229, 238

Washington, George
 president, 79, 219

WATCH, xv, 1, 106, 107, 108, 109, 111, 113, 134, 154, 166, 173, 238

water, xiii, xvi, 2, 3, 4, 5, 6, 14, 15, 17, 18, 19, 20, 21, 22, 23, 24, 25, 26, 28, 29, 30, 31, 34, 35, 36, 41, 43, 44, 45, 48, 49, 53, 54, 55, 57, 59, 62, 63, 65, 70, 73, 78, 82, 83, 87, 88, 91, 93, 95, 96, 97, 98, 101, 106, 107, 110, 111, 116, 120, 124, 126, 129, 137, 140, 141, 147, 148, 149, 150, 151, 159, 160, 164, 165, 167, 172, 173, 180, 181, 191, 197, 198, 201, 203, 205, 206, 208, 210, 214, 215, 220, 222, 227, 228, 229, 230, 231, 232, 233, 234, 236, 237

water cycle, 147

water vapor, 2, 3, 4, 18, 19, 20, 21, 35, 49, 55, 57, 82, 83, 91, 95, 96, 97, 101, 124, 140, 141, 180, 181, 198, 214, 215, 227, 228, 230, 232, 234, 237

weather, iii, ix, x, xi, xii, xiii, xiv, xv, xvi, xviii, 1, 2, 3, 4, 5, 6, 14, 15, 16, 17, 20, 22, 24, 27, 28, 32, 35, 39, 40, 41, 42, 43, 45, 46, 47, 48, 49, 50, 51, 52, 53, 54, 56, 59, 60, 61, 62, 66, 67, 68, 72, 73, 75, 76, 77, 79, 85, 87, 88, 89, 90, 92, 94, 96, 97, 98, 99, 101, 102, 103, 105, 106, 107, 108, 109, 110, 112, 113, 114, 115, 116, 117, 118, 119, 120, 121, 122, 123, 124, 126, 127, 128, 129, 130, 132, 133, 134, 135, 136, 137, 138, 139, 140, 141, 142, 143, 144, 145, 146, 147, 148, 151, 152, 153, 154, 155, 156, 157, 158, 159, 160, 161, 162, 163, 166, 167, 175, 176, 177, 184, 186, 187, 189, 190, 192, 194, 195, 197, 199, 200, 201, 202, 203, 204, 205, 206, 209, 212, 213, 215, 216, 219, 220, 221, 223, 224, 225, 226, 227, 228, 230, 231, 232, 233, 234, 236, 238, 250

weather forecast, 113, 114, 115, 116, 118, 128, 133, 134, 136, 137, 139, 140, 142, 145, 152, 153, 154, 156, 167, 192, 199, 200, 201, 204, 231 weather forecasting, ix, xi, 114, 119, 132, 133, 136, 144, 167

weather prediction, 134, 152, 199, 202, 232

weather radio, 103, 166

Weather Research and Forecasting Model
    weather model, 138

West River
    South Dakota, 28, 49, 52, 82, 86, 110, 127, 149, 150, 173, 175, 185, 186, 190, 214

west trade winds, 26

wet adiabatic temperature lapse rate, 82

Wickersham, Justin
    NewsCenter1 Sports Director, 213

wildfires, 38, 39, 78, 89, 90, 99

wind, xi, xiv, xv, 4, 8, 15, 22, 23, 26, 27, 32, 33, 35, 37, 39, 41, 42, 49, 51, 61, 62, 63, 64, 66, 67, 68, 69, 70, 71, 72, 73, 74, 75, 77, 78, 79, 83, 91, 92, 98, 99, 100, 101, 104, 108, 113, 114, 117, 118, 119, 128, 129, 139, 140, 142, 145, 146, 147, 153, 155, 160, 173, 175, 179, 184, 186, 190, 204, 206, 207, 208, 209, 214, 219, 220, 222, 226, 227, 228, 229, 230, 233, 234, 235, 237, 238, 250

wind chill, 61, 62, 63, 64, 67, 72, 186, 207

winter, x, xiv, xv, 1, 4, 11, 21, 33, 34, 43, 44, 45, 52, 55, 64, 65, 70, 71, 72, 76, 77, 78, 80, 82, 85, 118, 119, 138, 141, 145, 159, 160, 161, 162, 164, 168, 169, 170, 171, 172, 174, 177, 178, 179, 180, 181, 182, 183, 184, 185, 186, 187, 189, 190, 191, 192, 193, 195, 198, 206, 209, 211, 221, 222

winter solstice, 11, 177, 178, 192, 193

Winter Storm Atlas
    2013 blizzard, 185

winter storms, x, 4, 52, 76, 85, 141, 186

Woods, Tiger
    golfer, 24, 161

Y

Y2K, viii, 11, 215, 216

Yates, Col. Don
    meteorologist, 200, 204

yule or old moon
    January moon, 11

Z

ZULU time, 218

🍃 ADDITIONAL INFORMATION

For specific information about the current and past weather in and around the Black Hills, South Dakota area, such as temperature, wind, severe weather or snow information, go to the Rapid City National Weather Service web site at https://www.weather.gov/unr/.

For general information about climate, weather forecasts, severe weather and the oceans nation-wide go to the National Oceanic and Atmospheric Administration at https://www.noaa.gov/.

For specific climate and historical data across the United States visit the National Centers for Environmental Information at https://www.ncdc.noaa.gov/.

For the latest information on atmospheric research visit the National Center for Atmospheric Research at https://ncar.ucar.edu/.

For specific information regarding tropical storms and hurricanes visit the National Hurricane Center at https://www.nhc.noaa.gov/.